Mathematical Relationships
in Education

Routledge Research in Education

1. Learning Communities in Education
Edited by John Retallick, Barry Cocklin and Kennece Coombe

2. Teachers and the State
International Perspectives
Mike Bottery and Nigel Wright

3. Education and Psychology in Interaction
Working with Uncertainty in Inter-Connected Fields
Brahm Norwich

4. Education, Social Justice and Inter-Agency Working
Joined up or Fractured Policy?
Sheila Riddell and Lyn Tett

5. Markets for Schooling
An Economic Analysis
Nick Adnett and Peter Davies

6. The Future of Physical Education
Building a New Pedagogy
Edited by Anthony Laker

7. Migration, Education and Change
Edited by Sigrid Luchtenberg

8. Manufacturing Citizenship
Education and Nationalism in Europe, South Asia and China
Edited by Véronique Bénéï

9. Spatial Theories of Education
Policy and Geography Matters
Edited by Kalervo N. Gulson and Colin Symes

10. Balancing Dilemmas in Assessment and Learning in Contemporary Education
Edited by Anton Havnes and Liz McDowell

11. Policy Discourses, Gender, and Education
Constructing Women's Status
Elizabeth J. Allan

12. Improving Teacher Education through Action Research
Edited by Ming-Fai Hui and David L. Grossman

13. The Politics of Structural Education Reform
Keith A. Nitta

14. Political Approaches to Educational Administration and Leadership
Edited by Eugenie A. Samier with Adam G. Stanley

15. Structure and Agency in the Neoliberal University
Edited by Joyce E. Canaan and Wesley Shumar

16. Postmodern Picturebooks
Play, Parody, and Self-Referentiality
Edited by Lawrence R. Sipe and
Sylvia Pantaleo

**17. Play, Creativity and
Digital Cultures**
Edited By Rebekah Willet,
Muriel Robinson and Jackie Marsh

**18. Education and Neoliberal
Globalization**
Carlos Alberto Torres

**19. Tracking Adult Literacy and
Numeracy Skills**
Findings in Longitudinal Research
Edited by Stephen Reder and
John Bynner

20. Emergent Computer Literacy
A Developmental Perspective
Helen Mele Robinson

**21. Participatory Learning in the
Early Years**
Research and Pedagogy
Edited by Donna Berthelsen, Jo
Brownlee, and Eva Johansson

**22. International Perspectives on
the Goals of Universal Basic and
Secondary Education**
Edited by Joel E. Cohen and
Martin B. Malin

**23. The Journey for Inclusive
Education in the Indian Sub-
Continent**
Mithu Alur and Michael Bach

**24. Traveller, Nomadic and
Migrant Education**
Edited by Patrick Alan Danaher, Máirín
Kenny and Judith Remy Leder

**25. Perspectives on Supported
Collaborative Teacher Inquiry**
Edited by David Slavit, Tamara
Holmlund Nelson and Anne Kennedy

**26. Mathematical Relationships in
Education**
Identities and Participation
Edited by Laura Black, Heather Mendick
and Yvette Solomon

Mathematical Relationships in Education

Identities and Participation

Edited by Laura Black, Heather Mendick and Yvette Solomon

Routledge
Taylor & Francis Group
New York   London

First published 2009
by Routledge
270 Madison Ave, New York, NY 10016

Simultaneously published in the UK
by Routledge
2 Park Square, Milton Park, Abingdon, Oxon OX14 4RN

Routledge is an imprint of the Taylor & Francis Group, an informa business

© 2009 Taylor & Francis

Typeset in Sabon by IBT Global.
Printed and bound in the United States of America on acid-free paper by IBT Global.

All rights reserved. No part of this book may be reprinted or reproduced or utilised in any form or by any electronic, mechanical, or other means, now known or hereafter invented, including photocopying and recording, or in any information storage or retrieval system, without permission in writing from the publishers.

Trademark Notice: Product or corporate names may be trademarks or registered trademarks, and are used only for identification and explanation without intent to infringe.

Library of Congress Cataloging-in-Publication Data

Mathematical relationships in education : identities and participation / edited by
  Laura Black, Heather Mendick, and Yvette Solomon.
    p. cm. — (Routledge research in education ; 26)
  "Book is the result of a seminar series entitled Mathematical Relationships: Identities and Participation held in the UK between September 2006 and July 2007"—Introd.
  Includes bibliographical references and index.
  1. Mathematics—Study and teaching—Congresses. I. Black, Laura. II. Mendick, Heather. III. Solomon, Yvette.
  QA11.A1M27635 2009
  510.71—dc22
                                                  2008053049

ISBN10: 0-415-99684-8 (hbk)
ISBN10: 0-203-87611-3 (ebk)

ISBN13: 978-0-415-99684-6 (hbk)
ISBN13: 978-0-203-87611-4 (ebk)

Contents

List of Figures   xi
Acknowledgments   xiii

1   Introduction   1
LAURA BLACK, HEATHER MENDICK, AND YVETTE SOLOMON

PART I
Selection and Assessment   5

2   Disabling Numbers: On the Secret Charm of Numberese and Why It Should Be Resisted   9
ANNA SFARD

3   Pain, Pleasure, and Power: Selecting and Assessing Defended Subjects   19
LAURA BLACK, HEATHER MENDICK, MELISSA RODD, AND YVETTE SOLOMON WITH MARGARET BROWN

4   Mathematical 'Ability' and Identity: A Sociocultural Perspective on Assessment and Selection   31
JEREMY HODGEN AND RACHEL MARKS

PART II
Choice   43

5   Telling Stories About Mathematics   47
MARK BOYLAN AND HILARY POVEY

6   Choice: Parents, Teachers, Children, and Ability Grouping in Mathematics   58
PETER WINBOURNE

7   Special Cases: Neoliberalism, Choice, and Mathematics   71
    HEATHER MENDICK, MARIE-PIERRE MOREAU, AND DEBBIE EPSTEIN

PART III
Curriculum   83

8   Appetite and Anxiety: The Mathematics Curriculum and
    Its Hidden Meanings   87
    JENNY SHAW

9   Questioning the Mathematics Curriculum: A Discursive
    Approach   97
    CANDIA MORGAN

10 The Role of Textbooks in the 'Figured Worlds' of English,
    French, and German Classrooms: A Comparative Perspective   107
    BIRGIT PEPIN

PART IV
Pedagogy   119

11 How Do Pedagogic Practices Impact on Learner Identities in
    Mathematics? A Psychoanalytically Framed Response   123
    TAMARA BIBBY

12 Hybridity of Maths and Peer Talk: Crazy Maths   136
    PAULINE DAVIS AND JULIAN WILLIAMS

13 Pedagogy, Discourse, and Identity   147
    STEPHEN LERMAN

PART V
Teacher Development   157

14 Mathematics for Teaching: What Makes Us Want to?   161
    PAT DRAKE

| 15 | Developing Mathematics Teaching Through Collaborative Inquiry
BARBARA JAWORSKI | 173 |

| 16 | What Does a Discourse-Oriented Examination Have to Offer Teacher Development? The Problem With Primary Mathematics Teachers
TANSY HARDY | 185 |

PART VI
Endings   199

| 17 | Identity in Mathematics: Perspectives on Identity, Relationships, and Participation
PATRICIA GEORGE | 201 |

| 18 | Participating in Identities and Relationships in Mathematics Education
PAOLA VALERO | 213 |

References	227
Contributors	243
Index	247

List of Figures

15.1　Nested layers of inquiry in the LCM project.　　174

15.2　A research cycle in developmental design.　　178

Acknowledgments

This book, like most, took a long time, and many people helped along the way. Its beginnings were in a small ad hoc research discussion group which included Melissa Rodd, Margaret Brown, Sheila Macrae, and Judith MacBean, as well as the three of us. Mostly we found ourselves talking at length about our own and other people's relationships with mathematics—something which felt important to us. After a while, Sheila retired and Judith went on maternity leave, and so this group was whittled down to a hard core of five. We decided to open things up by organising a series of seminars. This was a massive collective effort, and we want to especially thank: Melissa and Margaret, our co-organisers of the seminars; the Economic and Social Research Council and the British Educational Research Association for providing the funding (and Maria Goulding for her support in securing the latter); Angela Kamara and Lindsay Melling, who administered the whole thing smoothly (or as smoothly as is possible in such endeavours); and the hosts at the various universities we visited—Tom Macintryre at the University of Edinburgh, Jan Laugharne at University of Wales Institute Cardiff, and Tansy Hardy at Sheffield Hallam University. The seminars depended hugely on those who took part—whether as speakers, group chairs, note takers, 'ordinary' participants, or some combination of these. Thanks to everyone who took time out of their busy schedules to participate. These events were pleasurable for us, both intellectually and socially, and we each feel we made many new friends through them. Finally, we appreciate the way that Leone Burton encouraged this book—she had wanted to publish it in her series on mathematics education before illness took over. Thanks to Routledge for taking it on in her place.

We each have some personal thanks . . .

Laura writes: I have many people to thank, both for their support and guidance and for continuing to foster my interests in learner identities within the field of mathematics education. This includes my colleagues on the TransMaths Project at the University of Manchester, and particularly Julian Williams for his continued support and guidance. However, I would especially like to thank my fellow editors and good friends, Heather and Yvette. During the course of completing this book, I experienced the life-changing event of giving birth to my first child, Thomas. I am eternally indebted to both Heather and Yvette for enabling me to stay involved in the book during this time and for taking on the extra work that this entailed. Additionally, my parents, Michael and Rebecca Black, and my husband, Stephen Foster, have provided much support and babysitting, for which I am very grateful.

Heather writes: I would like to thank the people with whom I have worked during the preparation of this manuscript, at both the Institute for Policy Studies in Education at London Metropolitan University and at Goldsmiths University of London. In particular, Anna Carlile, Anna Traianou, Katya Williams, Kim Allen, Marie-Pierre Moreau, Nicola Rollock, Rosalyn George, Sarah Smart, and Sumi Hollingworth have all helped make coming to work every day a little less painful. I have learned a lot about identities and mathematics from conversations in the Critical Mathematics Education Group and the Birmingham Science Education Research Group, from discussions with Anna Llewellyn and Cathy Smith whose doctoral research in this field I have been lucky enough to supervise, and from my collaboration on this book with Laura and Yvette. Outside of work, my mum, Ruth Mendick, my brother, Graham Mendick, and my friends Susan Shamash, Manoj Chaudhary, Farah Lalljee, Soraya Wasenius, and John Bayer have been important to me for lots of reasons.

Yvette writes: I am grateful to a number of people whose enthusiasm and ideas have had a major impact on my own understanding of mathematics and identity. I owe much to our original discussion group and the subsequent activities in and around the seminar series (thanks Heather for making this happen!). The learning carried on during the preparation of this book, with many challenging and insightful conversations around the preparation of our chapter with Melissa and Margaret, the fostering of new relationships with contributors, and of course the many exchanges with Heather and Laura which have been so fruitful. I would also like to thank the Manchester University sociocultural theory interest group for their generous inclusion of me in their many activities over the last few years. Finally, thanks go as ever to my family—John, Bridie, and Rosie O'Neill—for simply being there.

1 Introduction

Laura Black, Heather Mendick, and Yvette Solomon

This book is the result of a seminar series entitled *Mathematical Relationships: Identities and Participation* held in the UK between September 2006 and July 2007. The seminars were funded by the Economic and Social Research Council and the British Educational Research Association. We organised them along with Melissa Rodd and Margaret Brown and spent a year touring the UK, starting in London, then heading to Manchester, Edinburgh, Cardiff, and Sheffield, before returning to London for the final seminar. These six seminars brought together a number of researchers and practitioners working in the areas of mathematics education and in education and identity to discuss how we can better understand patterns of participation in mathematics. This book is an edited collection mostly of the papers presented at those seminars, with additional contributions from other international scholars.

The starting point for the series was the apparent paradox of better results in mathematics examinations juxtaposed with an increasing rejection of the subject at the postcompulsory level. This is an international concern, noted in Europe, North America, Australia, and the UK (Boaler, 2000b; Holton, 2001; Seymour & Hewitt, 1997; A. Smith, 2004). To make sense of this situation, we need to focus on the relationships that learners form with mathematics in the context of formal schooling. Central to these relationships is learners' developing sense of self and their understanding of the part played by mathematics in it. A focus on identity provides a means of understanding the processes that produce this 'paradox' and a way of beginning to address it by unpacking learners' emotional responses, their self-positioning with respect to others, and the surrounding discourses of mathematics education. Explorations of identity are therefore the central focus of this book.

Identity is a relatively new concept in mathematics education, although it is more widely used in educational and social research. The seminars reflected this fact, and we deliberately set out to include some researchers working in the field of identity who were not mathematics education researchers—this was a fruitful move, and the book correspondingly offers space for dialogue between this emerging exploration of identity in mathematics education and these wider theoretical perspectives.

Many disciplines deal with identity, and, within each of these, there are complex and overlapping ways of understanding it. The three perspectives on identity we selected for the series, and hence the book—sociocultural, discursive, and psychoanalytic—have interesting and contrasting things to say on how people relate to mathematics, focussing, respectively, on practices and participation, language and power, and the role of the unconscious. These theoretical perspectives are very broad, and the authors interpret them very differently. Here we offer only the briefest of orienting introductions. Sociocultural theories derive from psychology and anthropology (Holland, Lachicotte, Skinner, & Cain, 1998; Lave & Wenger, 1991; Wenger, 1998) and have been widely taken up in education (e.g., Bloomer & Hodkinson, 2000), including mathematics education (e.g., Boaler & Greeno, 2000). These approaches view identity as being co-constructed through participation in social practice and seek to understand the 'cultural models' agents use, and are positioned by, in their identity work. Discursive approaches view identity as the result of the subject's interpellation into discourse, systems of knowledge, and practice which construct objects; this process is inseparable from relations of power (Foucault, 1980). These approaches are of growing importance within the field of mathematics education (Lerman, 1998) as are psychoanalytic approaches (Evans, 2000) which define identity in terms of an interaction between conscious and unconscious processes and underline the value of focusing on the role of emotional and relational factors (Britzman, 1998). There are many tensions between these approaches. For example, there is a tension between the discourse insistence that we should not look inside people for explanations and the psychoanalytic concern with unconscious processes such as anxiety, fantasy, and defensive strategies. It is precisely these tensions that we feel make bringing these very different perspectives together so productive for addressing our overarching question: What can a focus on identities and relationships bring to understanding issues of inclusion in and exclusion from mathematics?

As we have said, in organising the series, and hence this book, we turned to the broad policy context of the 'paradox' of rising performance sitting alongside falling participation. We want to explore how policy is currently played out in and through mathematics education and the impact of policy on learner identities in relation to mathematics. Research indicates that the role of identity is most visible in formal contexts where learners are subject to policy-driven institutional structures which impose categorisations on them as 'good at' or 'not good at' mathematics via assessment and selection, pedagogy, and curriculum. In addition, the widespread emphasis in neoliberal policy on 'choice' presents a further context for (self-) definition. These themes, together with that of teacher development, provide our organising framework. We introduce them briefly here.

An increasing policy emphasis on assessment and selection has led to high levels of anxiety impacting on how learners construct their identities (Reay & Wiliam, 1999; Shaw, 1995). Within mathematics, discourses of ability and the associated practices of setting correlate with gender, class, and ethnicity and their related identities. For example, Gilborn and Youdell (2000) found that ethnic minority and working-class pupils are more likely to be allocated to lower sets regardless of prior attainment, while high-achieving girls in top sets are more likely to position themselves as having 'less right' to be there and to experience anxiety (Mendick, 2005a). The experience of assessment-related ability judgements has a long-term impact: university mathematics students continue to depend on positive test results for their identity confirmation (Rodd & Bartholomew, 2006; Solomon, 2007b). Correspondingly, self-selection or choice of whether and how to approach the study of mathematics is gendered (Mendick, 2005b, 2006), classed (Macrae & Maguire, 2002), and raced (Francis & Archer, 2005) and related to perceptions and self-perceptions of dis/abilities (Rodd, 2005). In the area of curriculum, the specialised language forms and strategies used in mathematics teaching also differentially favour some social groups in terms of class (Zevenbergen, 2000), cultural background (de Abreu & Cline, 2003), and gender (Walkerdine, 1998). Research on pedagogy indicates that 'traditional' teaching often results in differential participation (Boaler, 1997; Solomon, 2007a); high achievers and those from social backgrounds with greater cultural capital are most likely to join in positive interactions with teachers (L. Black, 2004a, 2004b; Jones & Myhill, 2004). Lower set pupils are likely to experience a 'polarised curriculum' (Boaler, Wiliam, & Brown, 2000) which not only limits exposure to mathematics but also obscures the underlying mathematical principles (Dowling, 1998). However, a wide body of research indicates that mathematics can be made more accessible in classrooms which encourage exploration, negotiation, and ownership of knowledge and their corresponding identity shifts (Boaler & Staples, 2008; Solomon, 2008). This has implications for teacher development, but also requires recognition that teachers' own mathematical identities affect and are affected by the curriculum and assessment context (T. Brown, 1997), and that their own, often troubled, relationships with mathematics impact on the ways that they interact with learners about the subject (Bibby, 2001).

The book has six parts. The first five reflect these policy and practice foci, each addressing a central question—about selection and assessment, choice, curriculum, pedagogy, and teacher development. These five themes are, as the previous literature shows, significant since they act as 'essential circuits' of education policy and practice (S. J. Ball, 1994). The treatment of each theme follows a uniform pattern, opening with a short editorial introduction contextualising our question and the responses which follow. The three subsequent chapters then offer replies from sociocultural, discursive, and psychoanalytic perspectives (although not necessarily in that order). Finally, the sixth part contains two chapters in which the authors, both international scholars, respond to the book as whole. We hope that by interacting with the ideas in these pages you will, like us, have a chance to reflect upon your own mathematical relationships and those of other people.

Part I

Selection and Assessment

In their answers to the question 'How does the concept of 'ability' operating within mechanisms of assessment and selection (by school and within school) impact on learner identities in relation to mathematics?', the chapters in this part focus on the ways in which assessment practices within mathematics education select students as 'able', 'competent', and 'normal'. Given that 'high-stakes' assessment plays a prominent role within current educational policy and practice in the UK, as in many other countries, it is perhaps unsurprising that we find it to be central in shaping many learners' relationships with mathematics and the mathematical identities they develop. In light of this, the chapters in this part all address a common theme: They seek to explore how people are measured and measure themselves in relation to mathematics.

Utilising a discursive approach, Anna Sfard (Chapter 2) presents an overview of the dominant macro discourse of quantification and highlights how quantitative assessments (by which we are all measured) permeate all walks of life. She notes that this may be particularly damaging within education, where numbers are secreted within hidden discourses and then become labels which identify actors rather than actions in a particular context, such that once a label is ascribed to a student via their performance in a given activity (e.g., a test or exam), that label becomes the intrinsic property of the student concerned, to be used by teachers or policymakers to differentiate between students. How the label came into being through the original act of performance then falls from view. A prime example of this can be found in the chapter by Jeremy Hodgen and Rachel Marks (Chapter 4), where primary school pupils are labelled (and label themselves) according to their levels within the English National Curriculum—labels which become reified markers, detached from their origins in the children's examination performance.

Anna goes on to note the potentially damaging effect of such quantified labelling in restricting the number of 'designated identities' (Sfard &

Prusak, 2005b) that a given student is able to take up in the future. She quite rightly points out that assessment outcomes (particularly summative assessment) can all too easily become statements about the student's future, and thus one or two 'local failures' can bring about automatic positioning at the lower end of the 'ability'/'competence' spectrum in the category of 'those who cannot'. Thus, in analysing the macro discourses of quantified assessment, she acknowledges its dominance in 'shaping the world around us' and calls for further questioning of what the numbers mean—what competencies underlie the labels we use? And are they the kinds of competencies we wish students to have?

Whilst Anna's chapter highlights the power of what she calls the discourse of 'numberese' in positioning people, the next chapter in this part builds on this by looking at how people are drawn towards the discourse of measurement and the labels it proffers. Adopting the psychoanalytic standpoint of Melanie Klein, Laura Black, Heather Mendick, Melissa Rodd, and Yvette Solomon with Margaret Brown (Chapter 3) present us with three 'mathematical biographies' which narrate the authors' turbulent and sometimes ambiguous relationships with mathematics. For them, discourses associated with assessment and selection are central to defining the pain and pleasure which frame our relationships with mathematics—such discourses offer positions which invoke our psychic defenses (against anxieties) in a myriad of ways. Thus, whereas Anna highlights the power of 'numberese', this chapter focuses on the psychic damage of such discourses when revealed through individual stories. This exercise is particularly illuminating since it underlines how discourses of assessment and measurement should not be considered as unambiguously problematic since, for some, they are also enabling—individuals can and frequently do use them to construct positive relationships with mathematics (e.g., through personal comparison or competition) in their 'identity work' (i.e., in defining who they are). Throughout the chapters in this part (and in this book), we see repeated examples of people using various assessment outcomes to measure or define their success. But as the accounts of Nikki, Rachel, and Zoë show, this is risky business since the threat of failure is ever present. Rachel's account in particular suggests a desire to compare her achievements with others, despite her knowledge of the ways in which mathematics is typically associated with processes of selection. Such comparisons carry the threat of failure, of course; this threat is powerful and perhaps even attractive—we are reminded of the 'frisson of danger' (Bibby, 2006) which appeals in doing such identity work. Therefore, one might argue that the psychic damage of 'numberese' is not simply brought about through the positions which assessment and selection discourses make available; rather, we may actively subscribe to, and work, these discourses as we negotiate our fluctuating relationships with mathematics.

Finally, the chapters in this part also address an important question—how might we change assessment practices within mathematics education in order to both challenge the potentially 'damaging' consequences brought about by labelling and measurement and reduce the threat of failure which we are so vulnerable to? This is essentially an issue about how we can avoid using assessment as a selection mechanism—to separate those who 'can' from those who 'cannot'—and how we can avoid the inevitable exclusions which stem from this. Here, Jeremy and Rachel's chapter is particularly relevant. Using both pupil and teacher interview data, they explore the commonality between pupils' and teachers' negative experiences of learning mathematics and note the role that summative assessment plays in positioning teachers so that they reproduce the same patterns of exclusion that they themselves experienced. In light of this, they call for a more relational approach to assessment, one which emphasises the development of relationships and learning rather than labelling. Anna makes a similar argument in her discussion of 'assessment for learning', with its emphasis on providing teachers and learners with detailed information on their strengths and weaknesses as a more effective pedagogic tool than the summative assessment which she rejects. However, Jeremy and Rachel go on to argue that we should not abandon summative assessment altogether since it is not quantified discourses per se which are stratifying but a lack of critical perspective amongst the teachers who use them. They suggest that we should instead encourage teachers to develop their assessment literacy and a critical understanding of the contexts in which such assessments occur. The chapter by Laura, Heather, Melissa, and Yvette with Margaret also suggests that abandoning summative assessment (and thereby decoupling assessment from selection) would not provide a complete resolution to the challenges and injustices many face in doing mathematics. We are still bound to personal comparisons even if we introduce practices which move away from the quantitative discourses enacted through 'high-stakes' summative assessment. Yet, they suggest, attempting to detach assessment from student selection and teacher accountability may be a step in the right direction in terms of developing more positive, 'pleasurable' relationships with the subject and challenging the highly valorised status of mathematics as a 'right or wrong' discipline.

2 Disabling Numbers
On the Secret Charm of Numberese and Why It Should Be Resisted

Anna Sfard

Testing and measurement have always been inextricable parts of the processes of teaching and learning, but never did they occupy as central a place in educational processes as they do today. This unprecedented assessment frenzy is just a particular manifestation of a more general phenomenon: of our present tendency for speaking in numbers about absolutely anything, whatever the nature of the things that are discussed. Numerical discourses—the **numberese,** as they are collectively called on these pages—and their educational uses are the focal theme of this chapter. After illustrating the claim about the present prominence of numberese, I take a critical look at the reasons and effects of its popularity: I locate those properties of numerical discourses that appeal to educational decision makers, analyse underlying discursive mechanisms, and argue that some of those features of numberese that make it decision makers' favourite are a mere byproduct of certain discursive devices, and are thus a matter of appearances rather than of any genuine asset. In some extreme cases, such delusions may cause tangible damage. In conclusion, I posit that immoderate, uncontrolled use of 'numberese' with reference to people and their actions is detrimental rather than helpful to the educational enterprise.

THE HEGEMONY OF NUMBERESE

The point of departure for the claims I wish to make in this chapter is the general assumption that one cannot escape if one takes seriously Vygotsky's (1987) insights on the co-development of language and thinking and Wittgenstein's (1953) criticism of the thinking–language dichotomy: Human thinking is the individualised form of interpersonal communication, and the way we communicate with others and with ourselves has therefore a major impact on how we act and how we perceive the world (Sfard, 2007b, 2008; cf. Edwards & Potter, 1992; Harré & Gillett, 1995). Various types of communication, or simply discourses, differ from one another in their vocabularies and word use, in their mediators and routines, and in the narratives they produce. In particular, participants of each discourse have their own ways of constructing, defending, and eventually endorsing stories about the world. Among the most prominent and influential of these narratives are those we call **identities**—stories about who we are, with whom we belong, and what position we occupy among those who constitute our human environment. When discourses change, the whole world changes, and our identities change with it. In particular, the human world described in terms of numbers is definitely not the same as the world described in number-free language.

Although numerical talk has always been around, its current ubiquity is without precedent. Our discourses, whether spoken or written, are saturated with quantitative expressions. Numerical symbols are an inextricable element of any artefact and of any public space, and numerical talk can be heard whenever people open their mouths. Always part and parcel of discourses of natural sciences and economy—these discourses would simply not exist without the language of quantities—the numberese may not be an immediately obvious choice when it comes to humans. Still, we do employ quantitative terms also while talking about people, and we do so whether we describe individuals or collectives, and whether we refer to persons' physical properties or to the qualities of their actions. At a closer look, numbers are the principal ingredient of our identities: In addition to our bodily dimensions (which attract much more attention these days than they did in the past), we identify ourselves by test scores, exam results, final grades, intelligence quotients, ranks and levels, income, socioeconomic indexes, states of possession, risks of genetic maladies, sports records, energy levels, placements on waiting lists, popularity ratings, and so on. Because our capacity for measuring and our quantitative imagination are unbounded, there is no reason why this litany should ever end.

Add to this the omnipresent measuring tools that became available with the advent of information technology, and everything about us becomes quantifiable. Just to make my point, let me pause for a moment and take a quick look around. . . . I enter the Internet, and, at random, I find the following piece of news:

> [An economist], with her students as research assistants, staked out eight coffee shops in the Boston area and watched how long it took men and women to be served. Her conclusion: Men get their coffee 20 seconds earlier than do women. (There is also evidence that blacks wait longer than whites, the young wait longer than the old, and the ugly wait longer than the beautiful. . . .) (Harford, 2007)

In my e-mail inbox, I then spot a questionnaire with the help of which a group of researchers whose names sound vaguely familiar is trying to find out what should be considered as necessary ingredients of mathematics teacher preparation. 'We request no more than fifteen minutes of your time', they say in the letter. And, indeed, they are using a Likert scale, so all I have to do is to tick numbers in boxes (and nobody seems bothered about the time I may need to think about my choices!). When I shift my eyes to the desk, I see a pile of student papers waiting for grading. I then notice a request for recommendation that arrived a few days earlier. A colleague is up for promotion, and I am asked to assess his 'productivity level', the 'impact factor' of the journals in which he publishes his work, and his 'professional standing in comparison to scholars at a similar stage in their careers'. Eager to do my work as a referee on time, I am tempted to look at his *GoogleScholar* scores. Another tool to be found on Internet, called *Publish or Perish*, allows me, in a matter of seconds, to extract several additional measures: cites/year, cites/paper, cites/author, papers/author, and so on. Above all, I can now help myself with new indices—h-index, g-index, hc-index, hI-index, and several others, all of them touted as being directly indicative of the quality of scholars' work. Although I am not sure of how the new indices are calculated and why they are supposed to reflect what people have in mind when speaking about quality, I now have an instrument with which to compare the candidate to everybody else in academia. There are many other measures I can use: this person's grants, his scores in student surveys, the number of invited talks and keynotes he gave at different conferences—and the list is still long. Even if my instinct is to shy away from all these numbers, I do keep in mind that whatever I am going to say stands a better chance to be found truly convincing if I express it in numberese. After all, whoever knows a thing or two about the politics of discourses must realise that numerically grounded statements are not anything one can argue with. They have the power to overwrite any other.

The steep rise in the popularity and dominance of numerical talk is explicable in view of the current proliferation of number-producing devices. The Internet alone is an inexhaustible supplier of brand-new numerical measures, only some of which constitute an answer to a preexisting need for information and most of which are a mere byproduct of the very possibility of measurement. Many of the things that are now accorded with numerical labels were never before considered as in any way number-related. As will be instantiated later, some of the objects implied in the measures did not even exist before the numerical epithets brought them into being. But even if stronger than ever, our tendency for numerical labelling is not new. I am now going to argue that the attractiveness of the numberese in any context, but in educational context in particular, stems, above all, from its four ostensible strengths: its informativeness, generality, rigor, and objectivity. These are the properties that give it an appearance of the best way of talking for those interested in quick and effective decision making.

EDUCATIONAL USES OF NUMBERESE: WHY, HOW, AND TO WHAT EFFECT

The excerpts in Box 1 are typical snippets of educational numberese, that is, of texts made possible by the (discursive) activity of quantitative educational assessment. I will use these excerpts to instantiate the four properties of this latter discourse and to substantiate my skepticism about its helpfulness. You are advised to examine the examples carefully before continuing reading.

Box 1 Examples of text produced as a result of quantitative educational assessment.

A. From a written testimony of University of Haifa undergraduate, recalling her experience as mathematics pupil in middle school:[1]

After studying with a friend for a whole week, hours after hours, I got the 'amazing' grade: 18 [out of 100]. That very day I told my teacher that this was the last time he had seen me in his classroom. But the teacher told me that if I only tried, I'd perhaps be able, at the end of the year, to get 100, and certainly not less than 80. I finished my mathematics studies at level 3 [the lowest], but with the grade 96.

B. From a job advertisement in *Work* supplement to the British daily *Guardian* (10 November 2007, p. 24)

To qualify for the scheme, you'll need a minimum 2.2 degree in any discipline.

C. From *IEA's TIMSS 2003 International Report on Achievement in the Mathematics Cognitive Domains* (IEA, 2005, p. 23)

At the eighth grade, led by Singapore, 24 countries and the four benchmarking participants had achievement in the applying domain significantly higher than the international average. Romania, Bulgaria, Norway, and Serbia performed no differently than the international average and 18 countries performed significantly below this average. At the fourth grade, also led by Singapore, 14 countries and the US state of Indiana had achievement significantly higher than the international average, two countries (Italy and Australia) and the two Canadian provinces had achievement similar to the international average, and 9 countries had achievement below it.

D. From *Review*, the supplement to the British daily *Guardian* (Laity, 2007, p. 11)

Mary Beard vividly remembers a day in her first year at Newnham College, Cambridge, when one of her friends saw a marked essay lying on her desk. He picked it up and read the tutor's comment: 'This is very good; I think it would get the first.' 'You,' he spluttered, 'get a first?' 'Even in the mid-70s,' Beard recalls, there were 'lots of men who thought that women were destined only to get 2.1s. . . 'From that moment,' she laughs, 'I was bloody determined to show them.' And she has shown them: Beard is now a professor in Cambridge and the best known classicist in Britain.

Informativeness

We may be all too familiar with the kind of texts presented in the box to notice how tightly packed with information they are. Just think about the complexity hiding beneath the simple idea of grade. Take, for instance, the 18 and 96 in Excerpt A, the 2.2 in Excerpt B, or the 'first' and 2.1 in Excerpt D. All these numbers are but tips of icebergs; they are all end products of intricate discursive processes of differing length and depth: 18 and the 'first' have been produced through the evaluation of a particular piece of student work, 96 summarises several basic evaluations of this earlier type, and the 2.2 in the job advertisement refers to the final assessment, supposedly combining all the grades accorded to the student through numerous years of study. Be they as stuffed with a hidden discourse as they are, individual grades are nowhere near in their tacit complexity to the *international average* in Excerpt C. This time, hundreds of thousands, if not millions, of individual grades for individual test items had to be successively produced before they could be combined into test grades; the test grades were then brought together to yield national scores and averages; to arrive at the international average, this last set of numerical data had to be subjected to yet another intricate, multistep discursive procedure.

The single number in Excerpt C that epitomises uncountable discursive actions of thousands of students, evaluators, and data analysts is a product of repetitive reifying—of replacing lengthy texts (e.g., student's written answers to test questions) with numbers, applying new discursive processes to these numbers (as in the process of calculating the total grade for the test), replacing the new processes with new numbers, and so forth. Halliday (1987) calls the texts resulting from such recurrent reifications *synoptic*. In such texts, 'the world is a world of things, rather than of happenings; of product, rather than of process; of being rather than becoming' (pp. 146–147).

The mechanism of reifying is invaluable in scientific discourse, where it is used to describe the functioning of agentless objects, but it may be less than helpful when applied to people and their actions. Here, the price of thus earned communicational thriftiness and intensity may be too high to afford. First, the repetitive reifications lower the resolution of the resulting picture, often to such a degree that it becomes impossible to translate the assessment into a truly useful practical advice. Indeed, a single grade ties together thousands upon thousands of individual activities. In the numerical images of reality, people are only as different as their numerical labels allow them to be. The users of the numerical information are thus habitually overlooking the potentially significant differences hiding beneath uniform numerical surfaces. Second, in successive reifications, the discursive history of the numbers is irreversibly lost, and unpacking the grades into what has actually been done by individual students and teachers in specific situations becomes virtually impossible. In view of this, trying to interpret the numerical scores is a hazardous enterprise. How to account for Singapore's consistently high scores? Are they a matter of teaching methods, genetic predispositions of the students, or some cultural idiosyncrasies, which make the success irreproducible in other cultures (cf. Sfard & Prusak, 2005b; Wang & Lin, 2005)? Besides, did this success involve competences that we really wish our students to have? After all, the related processes of using mathematics in solving problems and responding to math exam questions, although deceptively similar, are in fact two different types of discursive activity, oriented towards different goals, stimulated by different prompts, and constituted by differently made choices (Lave, 1988).

Generality

Another apparent reason for the attractiveness of numerical assessments is their wide applicability. Once an action or a series of actions is evaluated and accorded with a grade, we tend to see this numerical label as indicative of a pretty general property of the actor. Think, for example, about the requirement of 'a minimum 2.2 degree in **any discipline**' in the job advertisement in Box 1 (Excerpt B, italics added). Such requirement is only justified in the mouth of a person who considers the grade as independent of specificities of the learning process and who believes that a satisfactory score in one subject is a reliable predictor of satisfactory performance in any other type of activity. To put it differently, the advertiser treats the grade as a general property of a person, rather than of the way this person has performed some specific tasks.

Although this kind of assumption seems empirically testable, we rarely feel any need for justification while letting it guide our decisions. Once again, certain discursive mechanisms may be responsible, at least in part, for this state of affairs. Consider the following sentence from a recent issue of *TIME* Magazine:

> The average American child watches four hours of TV a day. (*TIME* magazine, 2007, p. 10)

At first sight, nothing seems strange here. On closer look, however, one realises that the adjective *average*, which should be used as a descriptor **of a certain number** related to children's activities (one should rather say 'American children watch TV, on average, for 4 hours a day'), shifted to the noun **child** itself, thus creating a new entity, 'an average child'. The relocation of the term **average** implies the assumption that doing things in an 'average way' comes in clusters: You watch TV the number of hours equal to the national average of watching hours—and you do all other things in an 'average' way. Whereas not directly related to educational assessment, this example provides a perfect instantiation of one of the many linguistic transformations that underlie our belief in the generality of messages carried by numerical labels and our resulting tendency for overgeneralising these messages.

Once translated from a property of an action into a property of an actor, numerical evaluations are likely to cause temporal overgeneralisation as well. Numbers accorded to a person for a particular performance stay the same as the performance goes on, and they are thus only too likely to act as self-fulfilling prophecies. Grades and the resulting epithets such as **gifted** or **learning disabled**, although constructed on the basis of one's former actions, are usually read as statements about the subject's future. As a result, low scores for past activities start perpetuating failure, whereas high grades are only too likely to beget further series of high grades. To see how this works, it is enough to look again at Excerpt A in Box 1. The bad grade, 18 out of 100, would have acted as a trigger for a downfall, if not for a prompt preventive action on the part of her teacher. (Note, however, that whereas this action dissuaded the student from giving up entirely, it did not prevent her from aiming low!) Not all the students, however, are blessed with this kind of support. Only too often do I see in my studies how one or two local failures lead to a sudden avalanche and an irreversible damage to the student's further mathematical learning (see e.g., Ben-Yehuda, Lavy, Linchevski, & Sfard, 2005). Let me stress that in the present context, the term **failure** does not refer just to one's inability to cope with certain types of problems, but also to the formal conferral of labels that automatically position the person at the very lowest end of the spectrum, in the category of 'those who cannot'. Note that this comparative aspect, made possible by the use of numbers, also played a decisive role in Mary Beard's case (Excerpt 4). Indeed, had the tutor satisfied himself with a qualitative, non-comparative appraisal, one that would have not implied her being *above* many of her classmates, most of the boys included, the effect would probably not be as dramatic as it was.

In summary, numerical labels tend to stretch beyond the sets of things for which they were originally designed, thus dangerously extending claims about samenesses, equivalences, and permanence. Generally believed to provide the best tools we have for describing reality, educational numberese is, in fact, the most underrated instrument for shaping the world around us. As long as we remain unaware of this latter fact, we mould reality with numbers unwittingly and not necessarily the way we would like this reality to be.

Rigour

Numerical labels divide the world into clear-cut, nonoverlapping categories, and the resulting picture seems neat, precise, and free from uncertainty. In the discourse of numerical assessment, there is little room for hedges and qualifiers and for modality other than full certainty: Either one has a certain grade or one hasn't; being in the 'upper 10% of the class' or having a

2.2 degree is a yes-or-no affair. Once a grade was accorded, the owner of the grade—be it an individual student, as in Excerpts A or D, or a country, as in Excerpt C—has been positioned in an unequivocal manner among all the other grade owners.[2] The neatness of numbers and their relations is bequeathed to the putative entities represented by the numbers. Thus, for example, once we start measuring 'mathematical competency', we are also likely to declare that the student who scored 70 on the mathematics test is 'less competent' or even 'less able' than the one who scored 85. We rarely ask ourselves what mathematical competency *is* apart from the numbers with which it is measured.

And yet, when we do think about the nature of measured entities—learning, competence, understanding, intelligence, and so forth—we realise that the reality is much messier than its numerical image. Dazzled by the algorithmic rigor of numerical calculations, we tend to forget that according grades to students' actions is a matter of imperfect, nonalgorithmic human judgment. Indeed, the clean numerical picture of students' learning may be but a cover for messiness. Only too often, the discourse of quantitative assessment forces unruly, ill-defined segments of reality into the straightjacket of crystal-clear numerical categories the way Cinderella's stepsisters pushed their big feet into the glass shoe: by cutting off whatever does not fit into the slot. Moreover, there is little consistency in how we perform this cutting from one situation to another. This is certainly not the way to attain rigorous understanding of what is going on when students learn, nor is it a good way to collect reliable insights on which to base improvements of our own teaching practices.

Objectivity

All that has been said so far does not yet explain why we fall so easily a prey to the temptations of quantitative assessment. True, such features as informativeness, generality, and rigorousness are all highly attractive, but what is the use of a discourse which only **pretends** to have all these advantages? The most obvious explanation for our weakness for numberese is that we are mostly unaware of the delusion. We take numbers bona fide, grateful for all they seem to be offering.

This confidence in numbers stems from yet another property of numerical statements: their appearance as mind-independent, world-imposed truths. Just look at the examples in Box 1: All of them speak about numerical assessments, but none of them mentions the assessor. The numbers and their interpretations are quoted as if they had no author. For instance, in Excerpt C, one reads that 14 countries have 'performed below [international] average', rather than being told that 'assessors accorded numbers lower than international average to 14 countries'. The matter-of-fact tone of the direct statement about 'below average performance' implies that international average and the national grades are 'objective'—they come from the world itself. As such, it is not up to humans to question the numbers or to try to change them. Similarly, in Excerpt 4, people are said to be 'destined' to the grades they get, and even if stated with tongue in cheek, this assertion would not have resonated with Guardian readers if it didn't build on some 'genuine' common belief in the direct relation between human beings and numbers.

Obliteration of the assessor is a highly consequential move. It results in what Bakhtin (1986) called monological discourse, a discourse whose narratives appear to be told in a single nonhuman voice—' 'the voice of the life itself,' 'the voice of nature,' 'the voice of people,' 'the voice of God,' and so forth' (p. 163). In monologically told stories, no space is left for human agency—monological storytellers do not seem to consider the possibility that different people might have produced different numerical measures for the same phenomena. Unmediated by human beings, the relation between numbers and the things they measure becomes a part of the external reality. In short, it is the discursively engineered appearance of mind independence and objectivity that fills us with confidence in numbers, blinds us to their pitfalls, and turns numberese into a powerful discourse, likely to overwrite any other.

LESSON: BEWARE OF EDUCATIONAL USES OF NUMBERESE

As argued in this chapter, all the apparent advantages of quantitative assessments—their ostensible informativeness, generality, rigor, and objectivity—are simply byproducts of certain discursive techniques. Through these techniques, diversity is successfully disguised as uniformity, and mere speculations acquire an appearance of unassailable truths. When combined together, these two delusions become infallible sources of unhelpful diagnoses and potentially disastrous educational decisions. Ironically, therefore, the favourite tool of those who purport to be helping students may, in fact, be one of the greatest obstacles to students' learning.

Once we become aware of this, we may also realise that we can, and probably should, argue with numbers and seek alternative ways of picturing reality. This, however, may be easier said than done. In principle, educational assessment does not have to be numerical. Take, for example, *assessment for learning* (see e.g., P. Black, Harrison, Lee, Marshall, & Wiliam, 2002; D. Cooper, 2006). If the aim is to give the student feedback about her progress, one obvious option is to analyse samples of her work, pointing to its specific strengths and weaknesses. In addition to its being free from the said weaknesses of numerical assessment, this type of reaction is probably much more effective: It provides the learner with specific, tailor-made information about what needs to be improved and how. However, non-numerical talk does not have the one special property of quantitative feedback that endears quantitative assessment to politicians and educational decision makers: It does not support easy comparisons, and it does not order all the learners in a single line, where everybody can be juxtaposed to everybody else. Without such linear ordering, there is no divide between **normal** and **abnormal** or **healthy** and **pathological,** and one cannot speak in terms **of red lines, averages,** and **tops** and **bottoms.** But all these latter expressions are exactly what decisions makers need to implement perhaps the most important of their tasks: to make selections and orchestrate exclusions (unfortunately, our society is not yet organised well enough to afford being left without any of these). Thus, one cannot expect them to give up their favourite tool without struggle.

Still, even if the task seems nearly impossible, we should probably continue opposing numerical assessment wherever we can. Educational numberese may be disabling rather than empowering the learner. This text should be read as an exhortation to disable it in return

NOTES

1. Collected in 1999 and translated from Hebrew by the author.
2. The language of positioning, by the way, is exactly the same whatever the nature of the graded entities: The expression 'x performs below average' is equally useful when x refers to an individual and when it denotes a whole nation (see Excerpt C again). The preservation of the language in passing from individual entities to encapsulated aggregates of such entities is what helps in camouflaging the quickly growing complexity.

3 Pain, Pleasure, and Power
Selecting and Assessing Defended Subjects

Laura Black, Heather Mendick, Melissa Rodd, and Yvette Solomon with Margaret Brown

In this chapter, we explore the use of psychoanalysis as a tool to gauge the impact of assessment and selection in mathematics on individuals in terms of their shifting 'internal' positioning in response to the experience of being judged and/or chosen, one that is central to an education in mathematics. In doing so, we present extracts taken from three of our 'mathematics biographies', narratives of our relationships with mathematics. We wrote them independently, circulating them only when all the stories were complete. When 'unveiled', they showed striking similarities in their presentation of ambivalences and their references to structural and personal comparison and competition, and to mathematics as a mediator of relationships with self and other/s. In this chapter, we make sense of them as accounts of defended subjects, selves 'forged out of unconscious defences against anxiety' (Hollway & Jefferson, 2000, p. 19), while also recognising the role of discourses in our defences. Following Wendy Hollway and Tony Jefferson (2000), we argue that 'subjects invest in discourses when these offer positions which provide protections against anxiety and therefore supports to identity' (p. 21). In an environment where selection and assessment are powerful markers of who we are, this psychosocial approach can give us new understandings of the pain, pleasure, and power in our relationships with mathematics, providing a critique of current practices.

THE DEFENDED SUBJECT

Our initial exploration of the idea of defence mechanisms with/in mathematics came about in response to Jaques Nimier's (1993) questionnaire-based research on high school students' responses to mathematics. Nimier categorised students according to a typology of defences. This consists of three 'phobic defences' against mathematics—where mathematics is the focus of projected anxieties and so becomes that which must be avoided, repressed, or projected out—and three 'manic defences' in which mathematics itself is used to reduce anxiety (e.g., by making an object, here mathematics, 'good' and investing in this 'good' by working on it). Using mathematics as manic defence involves the feeling that we are making, repairing, or constructing something.

Initially, Nimier's analysis resonated with our experiences. We could recognise within ourselves the spaces that doing mathematics fills and the fears that it engenders and quells. However, through writing our own mathematics biographies, we came to understand that, both methodologically and theoretically, assigning students to fixed positions within a typology of defence mechanisms is problematic—our stories were complex accounts of alternately approaching and avoiding mathematics, while the telling of them invoked defences once again. We came to see how Nimier's analysis implicitly drew on particular discursive positions, such as mathematics as absolute, rigorous, and ordered. In this and other ways, his explanations are complicit with dominant discourses of mathematics. His typology matched with parts of our narratives, but seemed to deny voice to other parts. In particular, we saw the ways that structures of assessment and selection are part of the conditions of possibility for our defences: this is an important social dimension absent from his work. Thus, through our analysis, we aim to understand our accounts as psychic **and** social, without reducing one to the other (Lucey, Melody, & Walkerdine, 2003).

The idea of selves as defended subjects is highly developed within the groundbreaking analytic work of Melanie Klein. Klein took the view that 'sufficient ego exists at birth to experience anxiety, use defence mechanisms and form primitive object-relations in phantasy and reality' (Segal, 1973, p. 24). In contrast with Freud's picture of human development as 'a matter of evolutionary progress from one psycho-sexual stage to the next' (Waddell, 2002, p. 2), Klein offers us a picture of the self as characterised by 'positions' or 'states of mind' which engender particular defences, anxieties, and types of relationship with others. For Klein, phantasy ('ph' rather than 'f' denoting its unconscious aspect) is central to all psychic processes:

> Phantasy emanates from within and imagines what is without, it offers an unconscious commentary on instinctual life and links feelings to objects and creates a new amalgam: the world of imagination. Through its ability to phantasize the baby tests out, primitively 'thinks' about, its experiences of inside and outside. (J. Mitchell, 1986, p. 23)

These phantasies are the means by which we make sense of the external world. They 'continue throughout development and accompany all activities; they never stop playing a great part in all mental life' (Klein, 1959, p. 251).

Klein theorised two particular states of mind which are distinct in shaping how we understand the world and 'self' in it: the paranoid-schizoid position (Klein, 1946) and the depressive position (Klein, 1935, 1940). Whilst each position can be associated with particular points in development as the baby learns to deal with separation and dependence on the mother or primary caregiver, they are neither biologically fixed nor maturational in nature. Throughout infancy and adulthood, we shift and slide between positions as we negotiate situations and relationships perceived as threatening, pleasurable, or both.

Our earliest relationships are characterised by the paranoid-schizoid position: 'It encompasses both the nature of the predominant anxiety, that is the fear of persecution, and the nature of defence against such fears' (Waddell, 2002, p. 6). Thus, the individual is portrayed as having an exclusive concern with their own interests: they may feel a sense of persecution in the face of pain or distress and engage in preservation of the ego at all costs. The baby is subject to pleasant and unpleasant external (and internal) experiences over which it is powerless. Pleasant experiences include being fed, held, and stroked, whereas unpleasant ones include waiting for a feed, being accidentally pinched with a nappy pin, and experiencing abandonment on waking. Alongside these are internal experiences (that may or may not coincide with the external events described previously) of fear, rage, despair, joy, and so on. So, projection and introjection come into play, and in these circumstances the mother/object becomes implicated in good and bad feelings. To this end, processes of projection (expelling) and introjection (taking in) are present from the start of life and underpin the interaction between the inner and outer worlds:

> And so he [sic] takes in the sense of having a bad mother. He has a bad mother within him. When she comforts and feeds him, and he has a good feeling, his mother again becomes good. He 'projects' his bad feeling and identifies her with it. He 'introjects' his experiences of her as calm, satisfying and good, and he himself acquires a good feeling within. He feels himself to be 'good'. (Waddell, 2002, p. 254)

This quotation illustrates this idea of 'splitting' which is central to the paranoid-schizoid position 'in which both people and events are experienced in very extreme terms, either as unrealistically wonderful (good) or as unrealistically terrible (bad)' (Waddell, 2002, p. 6). Splitting enables protection of the good object from the bad object, and along with idealisation and denial constitutes the defences that characterise the paranoid-schizoid position. Since we experience all objects, in part, through our phantasies about them:

> It is in phantasy that the infant splits the object and the self, but the effect of this phantasy is a very real one, because it leads to feelings and

> relations (and later on thought processes) being in fact cut off from one another. (Klein, 1946, p. 6)

As the typically developing infant grows, s/he takes on the depressive position, though the paranoid-schizoid position will be called on throughout life.

The move to the depressive position is believed to begin at about 3 months. At this point, the child starts to perceive and relate to their mother as a whole object rather than as part objects and so comes to recognise that both good and bad are contained within the same object. Feelings towards the mother are ambivalent since she is now understood to be the source of both gratification and frustration. Thus, the depressive position is associated with loss and guilt: there is a feeling of loss of the good object and of guilt at the realisation that previous aggressive impulses were directed towards it, since the splitting of good from bad object can no longer be maintained. Hence, 'the central task of the infant's elaboration of the depressive position is to establish in the core of his ego a sufficiently good and secure whole internal object' (Segal, 1973, p. 80). Reparation is a central defence associated with the depressive position:

> The drive to make reparation [...] can be regarded as a consequence of greater insight into psychic reality and of growing synthesis, for it shows a more realistic response to the feelings of grief, guilt and fear of loss [...] Since the drive to repair or protect the injured object paves the way for more satisfactory object-relations and sublimations, it in turn increases synthesis and contributes to the integration of the ego. (Klein, 1946, pp. 14–15)

Once again, we note that these two positions are not considered to be fixed states of mind which are purely developmental in nature, although they can account for changes in a child over time; as we will see, as adults we move between them depending on specific situations and relationships. 'States flicker and change with nuances of external forces and relationships—forever shifting between egoistic and altruistic tendencies' (Waddell, 2002, p. 9). It is in the shifts between positions that we can make sense of our relationships, including with/in mathematics.

MATHEMATICS AND THE DEFENDED SUBJECT

In this section, we use Klein's psychoanalytic approach as a 'lens' to read three of our mathematics biographies: those of Nikki, Rachel, and Zoë (pseudonyms). We focus on the ways that discourses of selection and assessment set the conditions of possibility for their phantasies about mathematics and so for their relationships with the subject. We look at how these play out in their shifting positions as defended subjects, as mathematics functions as both a sanctuary and a source of anxiety.

The Pains and Pleasures of Never Being Good Enough

Nikki, Rachel, and Zoë all have degrees in mathematics. Yet, despite being successful mathematically in conventional terms, they experience feelings of failure which point very firmly to just how at odds internal and external 'realities' can be and to why it is important to explore the distance between them in order to explore the impact of mathematics education assessment.

We begin with Nikki, who speaks of an early 'love' for mathematics. Although this predates her subjection to formal assessment processes, it is tied to her feeling that she is 'good at maths':

> I [can't] remember a time before I was good at maths; I just always was. Mum taught me sums and gave me Cuisenaire rods[1] to entertain myself with before I went to school and I remember that I couldn't see the point of unifix bricks when my infant school teacher tried to get me to use these.

Structures of assessment initially support this love by marking her out as 'able'. However, they later disturb this identification as they create increasingly fine distinctions and require constant renewal at ever-higher levels. Writing of her university examination results, Nikki says:

> I got my first [class degree] though it was a close one and so I didn't feel that I'd succeeded at maths. I got 5 out of my 14 questions out on one of the four papers. This paper was different, it was called the essay paper and was much more amenable to learning stuff and so I felt that I'd crossed the first barrier by learning stuff rather than doing the hard questions. My relationship with maths was never fully secure after that. I still loved it though and it still functioned as a safe space.

In an enormous contradiction, despite having a first-class mathematics degree from a prestigious university, Nikki has a well-rehearsed account of herself as someone who has not succeeded at mathematics. While it has provided a 'safe space' in the past, and continues to do so, its safety and the pleasure she gained from doing mathematics is not assured. In this extract, Nikki splits the object mathematics into 'good mathematics'—the unlearnable that earns a 'true' first—and 'bad mathematics'—which can be 'simply' learned and regurgitated. This is a splitting that draws on the gendered discursive opposition between 'real understanding' and 'rote learning' which constructs boys and men as being associated with the authentic creativity of the former and girls and women with the mindless rule following of the latter (Walkerdine, 1998). Perhaps it is significant that this coincides with Nikki's move from an all-girls school to a mixed university. Mathematics is, therefore, simultaneously a source of pains and pleasures, 'never fully secure' and a 'safe space', but these feelings gain expression through their contextualisation in assessment discourses. As we discuss in the next section, Nikki makes steps towards a tolerance of good and bad in the same object which is indicative of shifts in the direction of the depressive position in relation to mathematics.

Rachel, unlike Nikki or Zoë, began her part-time mathematics degree as a mature student while doing another job:

> I really like doing [the mathematics degree]. I have to work really quite hard in order to do it, because it takes up quite a lot of time, and I do it most weekends from February through to October, and I take it away on holiday, and I do it pretty well whenever I can.

This quotation gives a sense of Rachel's pleasure in mathematics which, as we discuss in the next section, she relates to reparation of her relationship with her father and to her sense of being challenged by mathematics. However, for Rachel, too, there is the pain of not being good enough which is tied to splitting between relational (good) and instrumental (bad) understanding:

> I'm not actually good at maths. This is because I never feel as though I really truly understand what I'm doing. I'm a very good rule follower, and I'm very good at keeping hold of something for a long time and not letting go until I've worked out what I need to do. And I really do think that a lot of my understanding is somewhat ritualistic in that I don't know how to apply it to new situations, so to me that means that I'm not really very good at maths. Well, I am good at it, but I don't have the depth of understanding that I would like to have.

This discursive contradiction ('I'm not actually good [...] well I am good [...]') indicates a splitting of the notion of what understanding might be (mirrored in longstanding debates on the nature of mathematical understanding). And while Rachel says she values understanding over instrumental performance, she experiences a painful attraction to an unrealisable performance goal:

> Unlike any other thing I've ever done, maths is the thing I want to score hundred percent in. I really want to get a first for this degree and I spend quite a lot of time working out how possible that is.

Once again discourses of 'ability' and who is 'able' at mathematics which are tied to formal assessments as indicators of effortless, all-or-none understanding are constitutive of Rachel's identification with mathematics, underpinning a phantasy of perfection and being able to 'please the mathematics'.

Zoë too describes mathematics as a source of pleasure and pain as she overcomes its challenges to gain ownership:

> I did not understand vectors. [. . .] But then [. . .] I 'got' the vector equation of a cylindrical helix; perhaps this was the first time I had made my own meaning of a mathematical concept rather than using the delivered one (integration did not become 'mine' until later). I love curves, surfaces, geometry though I have to work hard at it.

She writes of mathematics as a taken-for-granted secure space at school, something that 'happened without much thought'. She was taught at A level by a woman who was 'a completely reliable entity. Boring enough for maths to be comforting'. Zoë tells how she lost this comfort and with it her sense of self when, as a successful graduate, she changed universities:

> Fear came to maths at UniB. [. . .] I was less than nothing. [. . .] UniB was a different league from UniA—the thought of those first few analysis classes still brings out anxiety.

Her insecure positioning within this new and highly elite mathematics community forced her to increase her psychic investment which consequently made her vulnerable. Mathematics is no longer a defence against the world as it was for Zoë previously. In this new environment, she needs to defend herself **against** mathematics and the threat to survival, of death of the self ('I was less than nothing'), that this new league of mathematics presents. Given the contradiction with her identity as a graduate (and therefore successful) student, her relationship with mathematics needs to change. In her account, we see a description of a shift from mathematics as a place of safety to a site of persecution: at graduate school, 'I was so out of my comfort zone, so stretched by daily difficulties of doing it that hard maths stuff'. It appears that she splits mathematics into good and bad times, and, ultimately, as we discuss in the next section, she splits between mathematics and mathematics education.

Zoë, Rachel, and Nikki each experiences herself as inadequate in relation to mathematics. These feelings of inadequacy are related to the structures of assessment and selection in mathematics that create fresh exclusions at each new level. Both Zoë and Nikki are selected as 'good at mathematics' at an early age, and this positioning and the discourses that it draws on—mathematics as hard, abstract, and necessarily selective and therefore as challenging but 'natural' for some—initially support their identification with it. Both meet challenges as they continue their study of mathematics beyond compulsory education, and they rise to these. Initially, the 'harder' it gets, the more it works for Nikki, inscribing her within the discourse of the mathematically able as special. But ultimately, as for Zoë also, the harder it gets, the less useful it is as a defence. Eventually, they both split aspects of their identities from mathematics-for-its-own-sake. While Rachel comes to undergraduate study later in life, she too invests psychically in mathematics and enjoys the pleasures and pains of persistence in the face of difficulty. Ultimately, though, while these same discourses support identification with mathematics, they also create anxieties which must be defended against. For both Nikki and Rachel, assessment dominates their sense of inadequacy (only getting a 'close' first, not scoring 100%), and for Zoë, self-selection into the 'different league' of Uni B constructs a benchmark which she could not live up to. Therefore, for each of them, judgements of their mathematical abilities—official and unofficial, of self and other—create forms of anxiety. The defences and shifts of position that result are the focus of the next section.

The pleasures and pains of (not) doing mathematics

Taking up Nikki's story, by the end of her degree, she was left with a sense 'that doing any more maths was indulgent and pointless'. She became a teacher and began to develop a different relationship with the subject:

> I did my PGCE [one-year postgraduate teacher training qualification] and enjoyed doing the maths [. . .]. I guess that teaching maths shifted my relationship with it but I'm not sure how. I remember on my PGCE someone said that eventually your subject doesn't matter and you are just teaching children. [. . .] I didn't imagine myself making that transition when I heard it [. . .] I do think that I've made that transition now.

She offers an account of the development of her relationship with mathematics, starting with the neo-Platonic view of mathematics she held as an undergraduate:

> I thought [maths] was discovered rather than created, cos that was how it felt to me. This was a very powerful feeling and it was very hard to shift. It went in the first year of my [education] PhD. It was kind of like when I stopped believing in god. [. . .] I do experience it as amazingly freeing to have that different view of maths, now that I've got it. It's like maths can be a huge burden when it's like this thing external to you that you are enthralled/enslaved to.

Nikki starts by introjecting: associating herself as mathematical with her mathematics undergraduate identity and recalling how as a child she was nurtured with mathematics. Then she projects: seeing mathematics as dangerous with regard to her new identity. This change is also a change in the phantasies she has of mathematics. As a result of this shift, 'my sense of who I was and my place in the world shifted [. . .] I had to lose maths as special and to lose myself as special too, so that was painful'. This is not a total loss of self, however. Nikki replaces it with a new self, severing her connection with mathematics as she had seen it before, recasting mathematics, and using her 'growing academic identity and sociological understandings' to underpin this change.

The new relationship is even bound up with her physical sense of self, including a radical change of hairstyle. There is a growing ability to live in the depressive position signified by the acceptance of the existence of good and bad simultaneously in the same object which is evident in the juxtaposition of 'enthralled/enslaved' or even 'amazingly freeing' and 'enslaved to'. Nikki has come to a more stable projection of mathematics as dangerous, 'a huge burden'. However, this is associated with leaving mathematics, apart from 'Sudokus and the other odd bit'.

Nikki appears able to resolve the splitting of the once wholly good object by recasting her relationship with mathematics (and so herself) through her later academic (and, interestingly, competitive, audited, and so inordinately assessed) career. In contrast, Rachel tells us that she is unable to effect this change:

> Nikki sent me a questionnaire recently to try out, it asked whether you were doing a mathematical sciences degree or a maths degree, and my degree is called mathematical sciences and I was very keen to say to Nikki that that doesn't mean that the degree is applied rather than pure [. . .]. I was anxious that Nikki knew that my degree was just as good as a maths degree. [. . .] What I find really amazing about all this is the fact that I am somebody who thinks a lot about all of these issues and yet I'm unable to overcome them myself. So for instance one of the issues is the whole thing about ability and effort and speed. I know these are traps for the unknowing, and yet I fall into them every time.

In this extract, Rachel focuses on Nikki's perception of her, Rachel's, 'ability'. Nikki's own first, whilst failing to satisfy Nikki herself, does make her one of the 'able', in Rachel's eyes, and thus a point of comparison and a potential judge. Despite her claims about prioritising understanding, and her researcher knowledge which critiques the ways in which mathematics is discursively tied to processes of selection, she cannot save herself from the threat of failure through assessment and competition.

Rachel recognises the tension between her experience of intrinsic enjoyment of mathematics and her awareness of how doing mathematics can influence the way people see you: 'I'm impressed by people who do maths and of course the other aspect that is occurring to me of what all this means to me is that I wanted to impress people by being able to do maths myself'. She says this in the middle of writing at length about her relationship with her father:

> One thing which is true is that I have never been able to impress him with what I have done. When I was young I did an awful lot of things with my dad in order to please him, not to make him like me because that wasn't really an issue I think, but because I felt sorry for him somehow. So I used to go with him to watch motorways being built, which

> I really didn't want to do, and I used to do things with him because I think it made him happy. [. . .] But something that I've become aware of since I left home is that I really want him to say that I had done something well and he never really has. So I think that is a factor in why I do maths.

So we can read Rachel's relationship with mathematics, as she suggests, as also one in which reparation is sought in terms of her desire for a relationship with her father which includes mutual appreciation and support. We are forcefully reminded that the mathematical object related to is always a phantasied object and so 'can have energy invested in it in a very personal way' (Nimier, 1993, p. 32). Rachel works to restore the 'good object' in repairing her relationship with her father, invoking guilt and love as well as reparation. As we saw in the last section, Rachel speaks of how she loves doing mathematics, and works very hard at it, using all her free time, just as she gave up time for her father. This, combined with the dissatisfaction with anything other than a first, suggests that perhaps her work at mathematics and her desire for reparation are likely to remain unfulfilled. So, in contrast to Nikki's more persuasive resolution, we can see Rachel's story in terms of oscillation between a depressive state in which she effects an integration of good and bad in her (albeit reluctant) acknowledgment that she is 'good at maths' and a paranoid-schizoid position in which she cannot, and where the object of restoration—appreciation by her father—becomes a persecutor because it cannot be repaired.

Zoë resolves her own situation up to a point by taking up teaching: 'I started teaching which was wonderful: It gave me other ways of being with maths'. Analogous to the way that Rachel projects aspects of her relationship with her father onto her ability to do mathematics, Zoë projects aspects of her failed relationship with a male partner onto the subject. This is resolved through a splitting between mathematics and teaching mathematics. Mathematics as 'bad' was experienced as foreign and alienating, embodied in the mathematics PhD that she never completed ('his fault'); in contrast, mathematics teaching was experienced as 'good' and socially purposeful. The process of splitting is concerned, as always, with defending the self from the tension experienced doing mathematics: its simultaneous 'goodness'—using it to defend the self—and 'badness'—defending the self from it.

> No I did not get a maths doctorate. Probably why I had to compensate and do a maths ed[ucation] one.

Zoë's use of the word **compensate** here suggests that her doctorate was an attempt to make mathematics teaching equal to mathematics. However, that she feels this need further suggests that she still assesses the former in relation to the latter and so shifts and slides between the paranoid-schizoid and the depressive positions in her relationship to mathematics.

CONCLUSIONS: THE PSYCHIC POWER OF ASSESSMENT AND SELECTION

In this discussion of Zoë, Nikki, and Rachel's stories, we see some of the rich possibilities of applying psychoanalytic approaches to understanding mathematical identities and the ways that these are formed in relation to processes of selection and assessment. A psychoanalytic approach enables us to see how discourses of assessment and selection make possible particular phantasies about mathematics (and make others impossible) and that these phantasies provide us with ways to support our sense of self, but at the same time create anxieties which our selves need to defend against. In particular, it shows how we experience ambivalence towards mathematics understood in terms of shifts in defensive positions. As defended subjects, we frequently deal with our often painful/pleasurable relationships with mathematics and the institutions that deal it out to us by invoking splittings: in our stories, we oppose mathematics and mathematics education, 'good' mathematics and 'bad' mathematics. These splittings function as ways to defend against the anxiety that processes of selection and assessment provoke (Shaw, 1995) and draw on the various discursive positionings available in an environment where these processes are powerful social markers. For example, in the stories of Zoë, Nikki, and Rachel, we see discourses of 'specialness' (which are also explored by Heather Mendick, Marie-Pierre Moreau, and Debbie Epstein, Chapter 7, this book), discourses concerning a 'cachet of intelligence' (Nimier, 1993) associated with mathematics, and discourses of belonging, being accepted or rejected, and succeeding or failing. Although there are no simple matters of causality here, very often, as we saw, they contribute to people giving up mathematics.

One of the features of mathematics, when the discipline is considered itself as discursively constituted, is that there is an insistent opposition of right and wrong. Indeed, the clarity of the right/wrong dichotomy is reported by many postcompulsory participants in mathematics to be a reason for choosing to continue with the subject, as well as being a reason for nonparticipants to choose to quit mathematics (Mendick, 2006; Solomon, 2006). A question we can moot is whether attraction to this right/wrong aspect of mathematics is indicative of the persistence of the self in the paranoid-schizoid state of mind and thus the individual being potentially vulnerable to the potential catastrophe in moving from (being) right to (being) wrong. Perhaps this is why, especially in Zoë's and Rachel's accounts of selection and assessment, their stories are about how they use and are used by mathematics to create defences against failure **and** against success. Learners of mathematics at all ages and stages are likely to have their own unique ambivalent relationships with mathematics. The approach we have taken here might help develop teacher awareness of inevitable fluctuations and could provide justification for teachers to resist positioning learners too definitely—that is, to resist selecting as a result of assessment.

Selection and assessment are different, although they are both social processes connected in intricate ways. The current move to 'assessment for learning' (Black & Wiliam, 1998b), that has now been taken up by UK policymakers and practitioners, is a counter to summative assessments that are designed solely for selection and gate-keeping purposes. So would people still feel pain and pleasure from their own and an/other's assessment of their mathematics even if assessment was decoupled from selection or even if there were no formal assessment at all? Our analysis suggests that they would because eliminating formal assessments would not get rid of the need for reparation and other defences; however, it also suggests that their relationships with mathematics would, in many cases, be significantly and positively altered. However, reducing schools', teachers', and pupils' accountability in mathematics, at least in England, would be a major cultural shift. Assessment in mathematics is, in many settings, seen as high-stakes, whether as public examination or as weekly classroom test, and it provokes unconscious defences against mathematics. The highly valorised status of mathematics feeds the problem, while its critical 'filter' role compounds it. The detachment of mathematics from high-stakes assessment would perhaps also enable teachers and students to move away from the polarised right/wrong way of seeing the world and create possibilities for greater symbiosis between doing mathematics and working through the depressive position. This may not only enable more learners to retain pleasure in doing mathematics, but also help all learners to cope better with the uncertainties and ambiguities of life.

ACKNOWLEDGMENTS

We would like to thank Tamara Bibby and Helen Lucey for their generous and thoughtful readings of earlier drafts of this chapter and Jaques Nimier for providing our original inspiration.

NOTES

1. Cuisenaire rods and unifix bricks are referred to here. They are, respectively, a series of rods with lengths proportioned so that they can represent the numbers 1 to 10, with each number having a different colour, and a collection of multicoloured but otherwise identical small cubes that can be stuck together.

4 Mathematical 'Ability' and Identity

A Sociocultural Perspective on Assessment and Selection

Jeremy Hodgen and Rachel Marks

Mathematics is dominated by an approach to assessment and selection that conflates challenge and difficulty with notions of ability and discrimination. In secondary education[1], lower attaining students receive a largely remedial (and boring) curriculum, and most students regard themselves as weak mathematically. In this chapter, we explore how ability and assessment serve to produce and create largely negative forms of mathematical identity. We examine the continuities between teachers' and pupils' experiences of school mathematics and discuss how teachers' practices reproduce the same patterns of inclusion and exclusion to which the teachers themselves were subject. Finally, we suggest ways in which assessment could be reconceptualised.

The data that we draw on are largely from primary mathematics, examining the experiences of primary children and teachers. Much of the writing on ability and mathematics has focused on secondary education, and little attention has been given to how children's and teachers' earlier mathematics experiences contribute to this. But the emphasis on mathematical ability and setting in primary is increasing (e.g., Ofsted, 1998).

IDENTITY, ABILITY, AND LEARNING

In adopting a sociocultural approach, we focus on learning as a process of participation and enculturation (Kirshner, 2002; Sfard, 1998). Contrary to the dominant view of teaching and learning as transmission and acquisition, we see learning as identity development. Wenger's (1998) conception of identity as located in communities of practice is illuminating:

> Building an identity consists of negotiating the meanings of our experience of membership in social communities. The concept of identity serves as a pivot between the social and the individual, so that each can be talked about in terms of the other. It avoids a simplistic

individual-social dichotomy without doing away with the distinction. The resulting perspective is neither individualistic nor abstractly institutional or societal. It does justice to the lived experience of identity while recognizing its social character. (p. 145)

Wenger highlights the pivotal role for identity in understanding the relationship between human agency and social structure. Carr (2001) argues that learning is a negotiation between the individual learner and the social context in which 'a culturally and personally located social schema' may be 'transacted, redefined [...] resisted and, like discourse, called upon when the moment is opportune' (p. 527). In doing school mathematics, young people can adopt and adapt available learner and/or mathematical identities and thus become enculturated to the mathematical world whilst learning the explicit procedures and concepts outlined in curriculum documents. However, this identity 'choice' is both constrained and constraining culturally as well as individually. Carr's work demonstrates how this constrains individuals' learning orientations and, whilst some identities available allow for a positive enculturation, others exclude learners from mathematics. Identity is not only constructed through membership, but also in opposition to others (Gee, 1999). In Boaler and Greeno's (2000) study, many girls felt alienated by school mathematics because the only available identities were ones that required them to accept ideas without question.

We conceive of ability as an aspect of identity rather than an individual attribute. For example, M. Brown, P. Brown, and Bibby (2008) found that even relatively successful students, those predicted grade B at mathematics GCSE[2], regarded themselves as having failed at the subject in comparison to a perceived 'clever core' of fellow students, and, hence, were not choosing to study it post-16[3]. By 'imagining' their performance in comparison with other 'clever' students, these otherwise successful students 'learnt' to see themselves as lacking ability in mathematics.

Identities cut across the communities of practice a learner participates in:

> An identity is thus more than just a single trajectory; instead, it should be viewed as a nexus of multimembership. As such a nexus, identity is not a unity but neither is it simply fragmented. [...] Considering a person as having multiple identities would miss all the subtle ways in which our various forms of participation, no matter how distinct, can interact, influence each other and require co-ordination. (Wenger, 1998, p. 159)

Thus, a pupil's identity as a 'good school student' might be enacted in a variety of distinct communities, including but not exclusively the mathematics classroom. Alternatively, there may be contradictions and tensions between aspects of a learner's identity. Being a 'good English student', for example, might be at odds with being a 'good mathematics student', and being a 'good mathematics student' may be judged very differently depending on what mathematics set a student is placed in (Bartholomew, 2000). Indeed, the very notion of what counts as a positive school mathematics identity is itself contested (Barnes, 2000).

Confronted by tensions between the different aspects of their identities, individuals are compelled to negotiate and reconcile these different forms of participation and meaning in order to construct an identity that encompasses the membership of different communities. This process of identity reconciliation is central to an individual's ability to make connections and transfer meaning and knowledge between practices.

Holland, Lachicotte, Skinner, and Cain (1998) refer to two aspects of identity:

> We make an analytic distinction between aspects of identities that have to do with figured worlds—story lines, narrativity, generic characters, and desire—and aspects that have to do with one's position relative to socially identified others, one's sense of social place, and entitlement. These figurative and positional aspects of identity interrelate in myriad ways. (p. 125)

We find this distinction between a grounded positional identity and a figured, or imagined, identity useful. Whereas positional identity is grounded in specific communities and describes how people 'comprehend and enact their positions in the worlds in which they live' (Boaler & Greeno, 2000, p. 173), figured identity focuses on the ways in which individuals enact the less-localised identities, 'being a maths geek', for example. Holland et al. (1998) use the notion of co-development to emphasise the ways in which the learning or identity change takes place. They point to the space for human agency in improvising and authoring new practices and ideas. Thus, they conceive of individuals not only actively making sense of new situations, but also constructing new meaning in the process. But this improvisation occurs in response to the barriers inherent in these positioned and figured identities. So the creation of new ways of being occurs because of the constraints imposed by existing identities—both grounded and imagined.

METHODS AND CONTEXT

In this chapter, data are taken from two data sets: primary pupils and primary teachers. These demonstrate the role of relationships in identity development and represent our argument of reproduction in primary mathematics. The data focussing on primary pupils are drawn from Rachel's work on ability grouping in primary mathematics, together comprising 42 focal pupils in three primary schools: Riverside, Parkview, and Avenue[4]. The data for teachers focus on two cases drawn from Jeremy's study of primary teachers' professional development: Ursula and Alexandra.

Both studies adopted a similar methodological approach. The case studies of schools, teachers, and pupils were chosen as telling cases 'in which the particular circumstances [. . .] serve to make previously obscure theoretical relationships suddenly apparent' (J. C. Mitchell, 1984, p. 239). We wanted to understand the pupils' and teachers' mathematics as they themselves experienced it, and, thus, we used participant observation. Primary pupils were observed in their mathematics lessons and interviewed both individually and in groups to explore their constructions of ability and experiences of various grouping practices, whilst the teachers were asked to talk, in interviews, about their mathematical histories. Analysis was conducted using techniques drawn from grounded theory. Both studies specifically used constructivist-grounded theory, the more theory-driven approach developed by Charmaz (2000) in response to criticisms of grounded theory as narrowly empiricist and atheoretical. In this approach, in contrast to 'traditional' grounded theory, analytical concepts or categories are derived from reading the data alongside existing theoretical analyses. Rachel's study of pupils involved many more participants than Jeremy's study of teachers. So, with the pupil data, a greater emphasis was placed on the identification and use of Spradley's (1979) organising domains allowing sense to be made of the interrelations, connections, and potential multiple interpretations of events, whereas the analysis of the teachers' histories employed a narrative analysis in order to understand the teachers' trajectories of participation. We begin by exploring these two contexts separately before discussing how they may be working together in a cycle of identity construction and reproduction.

CONSTRUCTING IDENTITY IN PRIMARY MATHEMATICS

Ability and Mathematics Amongst Primary Children

We begin by looking at how identity, as we discussed theoretically, may be being developed through pupils' experiences in the primary mathematics classroom. Within this context, we present data that suggest how judgements and comparisons appear so pervasive that for pupils to talk about what mathematics is, is for them to talk about issues of assessment, ability, and labelling. Based on a study of Year 6 classrooms in England in the mid-1990s, Reay and Wiliam's (1999) study highlighted the influence of prolonged testing in which 'metonymic shift[s]' were experienced by pupils as they came to see themselves 'entirely in terms of the level to which [their] performance in the SATs is ascribed' (p. 346). It appears to be the case now that assessment is even more pervasive for primary pupils and is defining **who** the pupils are mathematically, rather than primarily either informing learning or recording attainment.

In England, pupils are regularly grouped according to ability for mathematics lessons. The pupils in the study reported here largely see themselves as the ability identifier (or National Curriculum level)[5] of their ability group, building their mathematical identities around these labels and levels: 'All the 3Bs go to the side of us, the 3Cs go in the middle and the 2As, they go to the end' (Y4 group interview, Riverside). Within classes, the labelling of table groups allows further discrimination in ascribed and developed identities: 'Table 1, that's clever, really really clever, table 2 is very clever, table 3 is very clever, number 4 is just clever. I'm on 1' (Peter, Year 4, Middle-Set, Riverside). The ease with which the pupils describe groups reflects findings in the literature of very little intergroup movement (MacIntyre & Ireson, 2002) allowing pupils to ascribe strong identities to each group and limiting the mathematical identities available to group members. Groups take on a unified characteristic; in school mathematics, groups are not composed of individuals, but of '3Cs'. Thus, ability group membership reinforces the effects of assessment in creating pupils' mathematical identities in terms of fixed trajectories of attainment.

Consistently, pupils talk about ranking. They have no difficulty in labelling how 'good' or 'bad' they are or of categorising other pupils into a dichotomy of the 'top' and the 'others'. Whilst all available evidence points towards a continuum of attainment (.i.e., there are no distinct high- or low-attaining groups) and shows that learning trajectories are not fixed (M. Brown, Askew, Hodgen, & Wiliam, 2009), the mathematical identities currently dominant in primary schools seem to be predicated on a notion that some are and some are not mathematical. The idea of the others, that some pupils 'ain't like us' or are only 'just clever', appears to feature strongly in how pupils think about themselves. Indeed, nearly every pupil spoken to would, without hesitation, reel off the names of multiple pupils who were 'bad at maths'. A repeated reification of labels/levels makes this seem the most natural of processes.

Rigid labelling appears to create a 'clever core', a top unattainable to the majority, similar to that found at the end of compulsory schooling (M. Brown et al., 2008). This 'clever core', who simply 'know lots of maths' (Lucy, Year 4, Middle-Set, Riverside), often without reason, appear at times to be viewed with a sense of awe: 'You know Mr Sherman's [top] set? I heard they're 4A' (George, Year 4, Middle-Set, Riverside). Thus, the mathematical identities available to the 'normal' student are thus constructed as lacking ability in comparison to the imagined 'other' of the clever core.

Mathematics, like the pupils themselves, is divided into two groups: hard and easy work. Those in top sets, and particularly the 'clever core', are identified as able to do:

> hard work ... [using] number lines, partitioning and working out ... word problems or something difficult like that ... really hard stuff ... like divide, times and subtract. (Y4 group interview, Riverside)

However, even for those in the 'clever core', mathematical identities are not static. Identity needs to be maintained, and there is an ever-present threat to one's identity (Gee, 1999). In the following example, two 'clever core' pupils and their teacher are faced with a dilemma:

> Mrs Woods (Y3, Riverside) gives each pair a pile of cubes for pupils to try to divide equally. Henry and Bill shorten the task by sharing out the cubes using a grouping strategy. They find they have 39 cubes so cannot share them out equally. Looking round the classroom they see it will be some time before the others have finished counting out their cubes one by one. Henry and Bill start to build with their cubes, 'making them into castles 'cause there's nothing else to do.' Mrs Woods asks for answers. Henry orders Bill to 'quick, put the cubes back into the tens', prodding Bill with his ruler. Bill seems annoyed and there is a fair amount of commotion. Mrs Woods ignores this. She asks who got the task 'right'. Henry seems keen to be right. He reaches under his neighbour's desk then immediately puts his hand up: 'Well we had 39, and then I found 1, so we had 40, which was 10 in each'.

Henry and Bill's efficient and mathematically interesting grouping strategy appears to be of little import. What matters is being 'right' **and** behaving appropriately. But the pupils' behaviour is at odds with the on-task behaviour expected of the clever core. To resolve this dilemma, all three (the pupils and the teacher) improvise (Holland et al., 1998) and act as if the off-task 'castle-building' never happened. For those perceived as 'clever', identity work such as this, co-constructed between the pupils and between the pupils and teacher in terms of what is and what is **not** drawn attention to, allows them to maintain their position/identity in relation to the norms of what it means to be good at maths: being 'right' and being 'on-task'. In maintaining a 'clever' positional identity, pupils imagine mathematics as never being 'wrong' and are hence discouraged from engaging mathematically:

> If you are quite clever in some way, sometimes you don't want to get something wrong because other people might say something about that, so I would rather not say anything. (Megan, Year 6, Set-One, Avenue Primary)

Those not included in the 'clever core' cover a vast spectrum of labels, yet their experiences and identity work share many similarities. Reified labelling allows pupils to think in a fairly derogatory manner about their peers: 'If you were only a 2A, you would be in Mr Young's [Bottom set]' (Y4 group interview, Riverside Primary). Unlike the mathematical identities ascribed to the clever core, the only discussion of mathematics that comes up in talking about others is in what they cannot do: 'They ain't like us, they ain't, they ain't like, they don't know the answers to 100 x 10, they won't know that' (Henry, Y3, Riverside). All the reasons that pupils give for pupils not being good at maths lie outside of the mathematics itself—for example 'it's because they shout out' (Jill, Y3, Riverside) or they are 'dyslexic' (Zachary, Y4, Avenue). Unlike the top pupils, lesson observations suggest that other pupils are regularly reprimanded for behavioural incidents which for the clever core are more likely, as in the previous Henry and Bill's case, to go unnoticed.

This is not to say that those pupils not labelled as 'able' do not experience positive moments in the classroom. However, these are often unrelated to the mathematics, with pupils being praised if they remember to put their hand up or for writing neatly. For these pupils, the possibility of developing a strong *mathematical* identity is severely limited. It is hardly surprising that many pupils eventually choose to identify themselves as nonmathematical. We note also one consequence of the dominant set of identities available to pupils appears to be that there is a limited engagement with the actual mathematics from *all* pupils. Those perceived as 'less able' are actively prevented from engaging, whilst the 'able', as we saw with Henry and Bill, seem so desperate not to lose their status that their energies are more substantially devoted to this, rather than to the mathematics. However, the pupils' struggles with their identities do not rest with them alone, but are seen to occur in co-construction with the teachers' actions, implicit and explicit. This is exemplified in this incident from a Year 6 (ages 10–11) top set mathematics lesson at Avenue Primary School:

> The teacher then asks the pupils a question. One boy quickly puts his hand up. The two higher-ability labelled pupils I am sitting next to let out exaggerated audible sniggers and in an animated conversation are quite derogatory of this boy, saying that 'he won't get it'. The teacher ignores this behaviour and the boy with his hand up, going to another labelled high-ability pupil for the answer.

Here, the teacher's behaviour implicitly dictates what is permitted and what is not. Such behaviours, whether realised by teachers' or not, may arise from their own relationships with and identities in mathematics, and hence lead to aspects of reproduction of ability labelling. We explore this possibility further now by turning to an analysis of the teachers' mathematical histories.

Looking Back on School Mathematics: Disconnection and Ability

We have shown previously how teachers are actively and passively engaged with pupils in the construction of mathematical identity and ability. We want to avoid blaming teachers. Indeed, teachers are themselves constrained by school structures (Horn, 2007). Rather, we want to explore how teachers come to see mathematics as 'ability' in order to understand how school mathematics could be constructed differently.

We note that Alexandra and Ursula were somewhat unusual primary teachers. They were amongst the first wave of numeracy consultants recruited to support and train teachers implementing the National Numeracy Strategy in England and Wales.[6] Hence, they were both positioned as primary mathematics 'experts'. Yet this positioning was somewhat uncomfortable because neither initially 'imagined' herself to be an expert in mathematics. Our analysis shows how the tension between these two aspects of identity, positioned and figured (Holland et al., 1998), enabled both teachers to reconstruct their mathematical identity more positively.

For Ursula, one particular incident crystallised her school mathematical experiences. She was in a small top set group who were taking a mathematics extension examination, an O/A, aimed at high attainers in mathematics at the same time as GCSE,[7] taught after school by her mathematics teacher, Miss Barker:

> I remember going from what I considered to be things that were really easy, drill and practice things, to this lesson where there was just this enormous algebraic equation going across the board. Absolutely enormous, and I'd walked in late because it was after school and I wasn't really over-keen on this, and I'd missed the first lesson, and I couldn't come to grips with this at all. [. . .] Supposed to work something out from it and everybody else in the room seemed to be able to do it except me. [. . .] The only good thing about it was I kind of looked around the room and thought—mm, this is an interesting group to be in. [. . .] [M]ost of them were the exceptional high fliers that just don't bear thinking about, you know, in your school year, they don't exist. They just get everything right, and you assume they do, and they don't exist as people.

Ursula's account has several resonances with the pupils' experiences described earlier. First, she categorised what she could do as 'really easy' common sense and what she could not do as incomprehensible (Coben, 2000). Thus, she constructs 'hard' mathematics as something separate and not related to her identity. Second, she identified those who can do this mathematics as not only 'exceptional' (as in the clever core), but also as different and alien, 'they don't exist as people' (Picker & Berry, 2000). Finally, she locates the source of the 'problem' individually: her own lateness and absence from lessons and 'I couldn't get it at all'. She went on to characterise her teacher as someone she didn't get on with and who was likely to make 'some derogatory comment' about her. Thus, her presentation is of herself as disconnected from mathematics as a discipline, from those who can do mathematics and from her mathematics teacher—and of a sudden, unexpected, and (to Ursula) almost incomprehensible transformation from being a mathematically 'able' student who could do mathematics to being someone 'who couldn't get to grips with it at all'.

Alexandra contrasted her relationships with two rather different mathematics teachers over several years:

> My maths teacher that I had [. . .] for four of the five years [of secondary school]. Miss Conway. [. . .] [S]he made it interesting and she made the links with, not exactly real life, but I can remember her teaching me compound interest by doing this whole thing about having a coconut ice factory [. . .] it was a silly context and that's what she thought eleven year old girls would be interested in [. . .] it made it a bit of fun [. . .] And she used to say things like—ooh, you are completely mad if you can draw a circle freehand on the blackboard. And of course she could draw a perfect circle. So although she was quite formal and you'd stand up when she walked in rooms, and . . . she was making a relationship with us [and] trust [. . .] in my third year I had Miss Morris, who was a graduate fresh out of somewhere wonderful [. . .] [She] had a really good degree [. . .] And that was a total waste of year for me. Because she couldn't teach. I mean, she had all the . . . she obviously had lots of maths knowledge [. . .] I just remember finding it hard to understand. [. . .] I got Miss Conway back again. But by then I'd lost some of my confidence in maths and when it came to calculus that was it. I just switched off.

Alexandra's account provides some contrasts with the pupils' experiences described earlier. In comparing her two secondary mathematics teachers—Miss Conway, who was 'making a relationship with us', and Miss Morris, who 'couldn't teach'—she highlighted several key features of Miss Conway's focus on relationships: the trust, interest, links, and making 'it a bit of fun'. In contrast, Miss Morris 'had a really good degree' and 'obviously had lots of maths knowledge', but 'didn't have very good control'. Alexandra located her difficulties with mathematics to the 1 year with Miss Morris, during which she found mathematics 'hard to understand' and 'lost some of my confidence in maths' and, despite getting 'Miss Conway back again', she 'just switched off' and disconnected from mathematics. But, crucially, Miss Conway's emphasis on relationships was not sufficient to overcome calculus. Alexandra's case suggests that a focus on relationships without these relationships being explicitly and deeply mathematical is on its own insufficient for the development of more positive mathematical relationships.

These experiences of school mathematics are very typical for primary teachers (Bibby, 1999). We describe elsewhere how both Ursula and Alexandra were able to reconnect with mathematics—for Ursula this was reconnecting with **being a mathematics teacher** (Hodgen and Askew, 2007) and for Alexandra this was reconnecting with mathematics *as a discipline* (Hodgen and Johnson, 2004). For Ursula, an extended course allowed her to revisit and challenge her experiences with Miss Barker and, thus, reimagine herself as a mathematics teacher, whilst Alexandra was able to question mathematics and, thus, develop an understanding of mathematics as a connected discipline (Askew, M. Brown, Rhodes, Johnson, & Wiliam, 1997).

Primary Mathematics, Identity, and Relationships

We have described how the processes of assessment (both formal and informal), coupled with perceptions of assessment, and the setting of children into different ability groups construct primary pupils as mathematically 'able' or not. Moreover, these inequitable processes are taking place at an increasingly young age. Mathematics dominated by external assessment and conflations with ability appears to influence the mathematical identities available to them. As pupils move towards the end of their primary school careers and secondary selection looms, any bridge between these two positions diminishes as imposed national and local assessment practices set what any one pupil can be. But primary teachers' mathematical identities too are constructed by 'ability' and assessment. This creates a vicious circle, with teachers replicating and reproducing the inequitable mathematics that they themselves experienced. The analysis of primary teachers' experiences suggests that reconstructing relationships—both between teachers and pupils and with mathematics—potentially offers a way of breaking out of this cycle.

DISCUSSION

In light of the evidence we have discussed in this chapter, we take issue with Anna Sfard's rejection of quantification (Chapter 1). The problem is not quantification *per se* in that *all* high-stakes assessments are prone to reification (Gardner, 2006). Assessments need to be manageable for teachers and pupils and, perhaps more significantly, they should open up rather than close down possibilities. Our analysis highlights the crucial roles of imagination and positioning in re-creating and reproducing (and thus reifying) existing structures that serve to include and exclude students mathematically. We emphasise that this is not something that is *done to* students. Rather, as our analysis indicates, agency plays a central role—students and teachers act together in constructing mathematical identities.

The problem is rather twofold. First, we believe that quantification is one key aspect of mathematical thinking. Understanding the strengths and limitations of quantification is key to being mathematical. Second, there is a lack of a critical perspective amongst teachers. One of us has recently conducted a collaborative study exploring how mathematics teachers could develop a fairer system of summative assessment involving teachers' judgements (P. Black, Harrison, Hodgen, Marshall, & Serret, 2007). Paradoxically, for a subject that includes measurement, the mathematics teachers felt any sampling procedures to be flawed because they might fail to capture some critical aspect of a students' learning. Ideally, *all* the mathematics teachers would have liked to be able to 'look over all students' shoulders all the time'. Making sense and enacting assessment requires a certain level of expertise on the part of teachers. Gardner (2007) describes this expertise as having three elements: assessment literacy, skills, and values. But this expertise is grounded in identity and is itself enacted within relationships between people. Therefore, we urgently need a more relational form of assessment—one that prioritises listening to pupils and promotes learning rather than labelling. Two current approaches offer possible ways of doing this. First, Boaler's (2008) relational equity approach demonstrates ways in which group assessment can be used to encourage pupils to take responsibility for each other's learning. Second, P. Black and Wiliam (1998a) have described how formative assessment can be used to inform learning. Yet, despite its apparent success, there remain unresolved issues and challenges with formative assessment. Its implementation by many teachers is superficial (P. Black et al., 2002) and mathematically weak (Watson, 2006). Indeed, really implementing formative assessment necessitates a transformation of the authority relationships that dominate mathematics classrooms. Yet without a greater focus on relationships, we believe that mathematics education is doomed to reproduce the negative and stigmatising conceptions of ability that we have described previously.

NOTES

1. In England, there are two phases of compulsory schooling: secondary (ages 11–16) and primary (ages 5–11).
2. The General Certificate of Secondary Education (GCSE) in England. GCSE examinations are generally taken at the end of compulsory schooling (age 16). Almost all students sit GCSE mathematics. In 2007, 34% of these students achieved a grade B or above.
3. Crucially, this perception is itself reinforced by teachers. This is vividly illustrated by the recent experiences of Jeremy's daughter. She is interested in studying Further Mathematics post-16. In every school that she has visited, the mathematics (and physics) teachers stand in stark contrast to teachers in every other subject. Whereas other teachers of other subjects invite her to come in and find out, all the mathematics (and physics) teachers have begun the conversation by saying that the subject is far too hard for most students, to which her reaction has been, 'Why do they **try** to put you off maths'?
4. Pseudonyms are used throughout for all names. These three schools represent a range of ability setting practices that pupils in England experience at Key Stage 2 (ages 7–11). At Parkview, setting is only introduced for mathematics and only in the final year (Year 6, ages 10–11), Riverside has recently undergone reorganisation with all pupils set for mathematics throughout Key Stage

2, whilst Avenue has a strong setting policy for mathematics, implemented whilst pupils are in Key Stage 1 (ages 5–7) with set positions predominantly retained across Key Stage 2.
5. Pupils in England and Wales are assessed against National Curriculum levels. Levels are subdivided into A, B, and C, with pupils expected to move up a complete level every 2 years. At the end of primary school (age 11), most pupils are expected to have attained Level 4.
6. The National Numeracy Strategy (NNS) was a national initiative in primary schools introduced across England in 1999/2000. One element was a systematic training programme organized by specially appointed Numeracy Consultants in every Local Authority.
7. The O/A is no longer available. Only a relatively small proportion of students took the O/A, and it was not available in all schools.

Part II
Choice

Perhaps more than the other parts in this volume, these contributions on choice show how our three theoretical lenses intertwine in addressing issues of mathematical identities. While each forefronts a particular lens in providing an answer to our question 'How do mechanisms of choice impact on learner identities in relation to mathematics?', they can all be seen as illustrative of the role of discourse, of figured worlds, and of psychic histories and emotions in their analyses of mathematical choices.

Working from the level of the individual, Mark Boylan and Hilary Povey (Chapter 5) explore the 'interior world' of one student teacher's choice *not* to study mathematics, reproducing her story without interjections as Louise tells how she experienced mathematics classes and how mathematics still has the power to force her back to childhood emotions, and to make her 'totally nine again'. In using Louise's story as their lens, they demonstrate the importance of individual histories and emotions in understanding relationships with mathematics and how these are shaped by wider beliefs and cultural practices. They suggest that Louise, while exercising choice in *how* she tells her story, has little choice in what that story is composed of. As Mark and Hilary point out, she is subject to compulsion in her experience of the institution of school and the power balance between teachers and pupils who are themselves part of a wider discursive positioning of the power of mathematics. We see these different strands combined in Louise's story as she flips between a reliving—in the present tense—of her emotions and a more distanced, past-tense, adult judgement which contextualises what happened within discourses of and about mathematics ability in her school and the figured world of its actors, the significance of their actions, and the extent of their agency. In reading Louise's story for ourselves, we can experience her pain and understand the power of it. We can also see how, both in her reliving and in her more retrospective account, Louise contemplates—but does not execute—resistance to the position she is offered. As the Holland et al.'s (1998) theory of figured worlds suggests, it is possible to refigure relationships if we are able to stand back from them and see them in perspective—so could Louise have refigured her relationship with mathematics? What resources would she need to draw on? While she tells us that the pressure of needing to perform well in mathematics made her 'angry, very angry', perhaps this emotion is not enough—it is only the adult Louise who has the resources to critique the power imbalance that she is victim of in her analysis of classroom processes, beliefs about mathematics, labelling, and the gendering of mathematics. The child Louise is, it seems, powerless to resist. As Hilary and Mark say, the storying of Louise shows the extent of psychic damage which has resulted from her experience of learning mathematics—as her story moves further into the present, we see how, despite her insights into her past, she is unable to refigure her relationship with mathematics: succumbing to the power of discourse and her own repeated positioning, she says 'maths is straight lines and I think in circles'. Ultimately, Louise blames herself for her failure—she is 'a bad workman (*sic*)'.

Peter Winbourne (Chapter 6) focuses on a sociocultural approach to choice, this time in the context of a school's choice to introduce ability grouping, and its effect on identities—not just of the children at the school, but of parents and teachers too. In this snapshot of the school at this crucial point in time, Peter shows that, in talking about the community of practice within which 'the choice' occurs, it is necessary to capture a sense of the wider context and discourses which frame what happens. Discourses of ability are at the forefront here, and their effect is very visible in what the children say. So, for example, we see how individual children position themselves within their school and its newly explicit organisation in accordance with the designated roles and actions arising from ability grouping and the hierarchical labels of 'accelerated', 'accelerating', and 'standard'. As Peter points out, the labels appear to engender a new 'sense of movement and pace' which clearly motivates some of the children and provides them with a positive identity space. The power of the groupings is also demonstrated by their effect on how children position each other—thus, Darrell is perceived with new respect due to his unexpected placing in the 'accelerating' group. There are potential pockets of resistance in this figured world, however: Aaliyah, who is in an 'accelerating' group and is proud to be 'gifted and talented', argues nevertheless that 'there must be gifted and talented in all different sets'. The children also express emotions associated with 'the choice': they talk about missing friends who are no longer in their class, and they also talk about loss—for at least one child 'standard' equals 'ordinary' and no more jokes with teachers, and for another it means not being able to go on a special 'gifted and talented' trip. But the new words ultimately appear to fix children in terms of their relative positions in the wider discourses of ability and speed which are hard to resist for children, parents, and teachers. As Peter concludes, it is difficult to 'think outside the box' of ability grouping.

Choice as a discourse in itself is the issue addressed by Heather Mendick, Marie-Pierre Moreau, and Debbie Epstein (Chapter 7). Their starting point is the centrality of consumer choice in the current sociopolitical climate—including its very high profile in education—and the apparently limitless choice that we have in deciding who we are. In what is in effect a context of *compulsory* choice, they address the issue of the discourse of subject choice and its assumption that the neoliberal self must be realised though self-actualisation, for which the individual is responsible. As they point out, this view ignores how choice is in fact constrained by structural factors such as gender, class, and race, and in so doing blames the individual for their own 'failings'. Their reports of young people's accounts of subject choice show once again the power of the discourse of 'ability' in self-positioning, while the frequency of the word 'always' suggests the force of repeated positionings and of designated identities in the figured world of mathematics classrooms. Thus, Dave, for example, recalls the classroom and his role in it of *always* being picked, *always* finishing first. Joanna too talks explicitly about how her teacher designated her as 'good at maths'. Relationships matter, though: recognition by another—the significant actor who is the teacher—is crucial. Also visible in these accounts are references to emotion—Dave and others who chose mathematics express strongly how it makes them *feel* good. Enjoyment is frequently mentioned, and those who chose it describe experiencing mathematics as 'an essential part of the self' which makes you feel 'different and special'. However, as Heather, Marie-Pierre, and Debbie point out, the very idea of 'specialness' is underpinned by the discourse of the selectivity of mathematics ability—it is simply not available to all. Supported by the process of othering, a desire to maintain oneself as 'other', and pervasive discourses of what constitutes 'real maths', the discourse of ability is shown in this chapter, as in the previous two, to have a lot to answer for: in all three chapters, we see how it is a major factor in undermining agency and choice.

5 Telling Stories About Mathematics

Mark Boylan and Hilary Povey

> We are part human, part stories. (Okri, 1983, p. 114)

When we think about issues of choice in mathematics education, we perhaps most readily focus on those particular moments where people make decisions, or decisions concerning them are made by others, that have observable outcomes. An obvious example is the choice made by learners to continue to study mathematics when the subject is no longer compulsory for them. However, sociocultural, discursive, and psychoanalytical enquiry (see e.g., Boaler & Greeno, 2000; T. Brown, Jones, & Bibby, 2004; Mendick, 2002; Solomon, 2007a; Walkerdine, 1988) all indicate that such moments of apparent decision are grounded in ongoing processes. In considering the issue of choice, we begin from a concern/interest in these processes, the experiences of maturation, of development and change, and the factors that are involved in this. In particular, we are concerned with how people themselves make sense of their experience of being in the world. This includes the relationship between a complex, phenomenologically interior world that frequently has hidden or unconscious, and invariably emotional, motivations and causes, and articulated human choice.

This making sense of the experience of being in the world we sometimes think of in terms of 'storying the self' (Povey & Angier, with Clarke, 2006) and the experience of identity as 'identity work' (Mendick, 2002). Such work occurs not only in relation to what is observable or exists in shared public space, but also and crucially in the interiority of what we experience as our personal self(ves). It is this interiority and the importance of 'meaning-in-life' (Yalom, 1980) which is the focus of the chapter and which provides the link to psychoanalytic concerns.

STORYING LOUISE

Our concern here is with Louise[1], a student who was coming to the end of a training course preparing to teach English in secondary schools[2]. We seek to re-present the ways in which she does and does not exercise choice in her mathematical life world. The chapter is woven around an extended research story, presented in the form of a monologue, and constructed from a semistructured taped interview with Louise. (Further details about the interview, research context, and construction of the monologue can be found in Boylan, 2004.) It is unusual to present a research participant's words at such length, but we adopt this form purposefully. The more usual practice of thematic selective quotation does not, in general, through the form of the text, evoke the wholeness and interconnectedness of the subjective life world which is our concern here.

The monologue also gives readers an opportunity, if they are willing, to try to stand in Louise's place: to understand 'understanding' as akin to being an understudy, that is, to be capable of standing in for Louise and perhaps, if involved in mathematics education, being willing to stand for Louise and for others marginalised and alienated by mathematics (i.e., to be engaged on their behalf for change).

We acknowledge the irony that, necessarily, the persona presented in the monologue is not our original participant nor is the monologue her story; rather, it is our story: Even in referring to her as Louise, we are already othering her (Fine, 1998). The monologue is a reconstruction and a creative interpretation of words she spoke in a taped interview. In the presentation of this research material as a monologue, we act as researchers who 'craft our text' (Povey & Angier, with Clarke, 2006, p. 461; but for a critique, see Stronach & MacLure, 1997) in a way that both makes it clear that it is a construction and creates, at least, the possibility to read some of the complexity of the stories that Louise chooses to tell about her self and her mathematics; we have attempted to retain polyvocality (Gergen, 1999)— not a single story about mathematics, but many stories. We stay close to Louise's words, but explicitly acknowledge the gap between experience and our research constructions of it.

Louise's story includes many themes that are familiar in literature on the experience and social practices of school mathematics; for example, disaffection, boredom, rote learning, confusion, gendered experience, labelling, desire for understanding, public and private shaming, and identifying and not identifying as 'good at maths' (Bartholomew, 2001; Boaler, 2000a; Boaler & Greeno, 2000; Boaler et al., 2000; Boylan, Lawton, & Povey, 2001; Mendick, 2003b; Nardi & Steward, 2003). However, these are only indirectly our concerns here. Rather, we are telling Louise's story because it speaks of and evokes a wide range of emotions and psychological states (e.g., anxiety, fear, hope, despair, envy, resentment, disappointment, anger, and shame). These are less often the focus of research attention, but are fundamental to and constitutive of the mathematical life worlds of learners.

The form of text that we offer requires different demands on the reader from other forms of academic text: It 'asks readers to engage in reading *as work*' (W.-M. Roth & McRobbie, 1999, pp. 505, italics original). Influenced by Moreno's psychodrama and action methods (Fox, 1987), we ask you to explore our use of the form of a monologue so as to 'role reverse' with Louise by actively reading her words aloud, with emotion, imagining if you can what it might be like to be in Louise's world, to have had her experiences, to try to feel the way she might. We ask you to engage actively in reinterpreting the text for yourself and to consider the extent to which Louise's relationship with mathematics is and is not a chosen one. We see compulsion and the absence of choice in the life world of child Louise. Teachers choose, mathematical objects exercise agency and act in and on the life world. However, we do see choice in adult Louise's retrospective imaginings, in her storied re-creations; but the choice is tempered by the psychic damage consequent upon the child's mathematical schooling.

LOUISE'S STORY

I was a good all rounder at subjects at junior school. However, I was very aware that I was not part of what my teacher, and I **hated** this term, called 'the maths pundits'. They would have extra work to do and I would just do mine. I cannot really remember very much, but I do remember Friday afternoons.

We used to have like a football league for the times tables. Your name was on a little strip and you used to get relegated to another division and you used to get put up. I was usually somewhere in the first division but only because I really, really tried. I was usually hanging around at the bottom of division one. I never learnt long division, for instance, because it wasn't worth teaching me that. My junior school teacher said that I'd never get it: 'There's no point teaching her it'. I don't think I was alone in that. I think it was only these maths pundits who got taught this long division.

Anyway it would be whoever's turn it was on the Friday afternoon. The two top from the first division, and then the next two, and it would go down, down this table of all the kids in the class. And you know you would be put up against somebody and you would both have to stand up. Everybody else is sitting down quiet listening to you, and you would both have to stand up and the teacher would ask you one of your times tables. And you had to be the first one in with the answer.

I **hated** it.

I remember that I used to stand up and my stomach would be going and I'd be going 'oh no I can't lose. . . can't do this'. You stood up, the two of you, in front of **everybody**; you stood up and fought it out.

I didn't like it at all. You couldn't get out of it and I used to **dread** it. I didn't worry about it all week but by the time it got to Friday morning and I knew it was coming Friday afternoon, yeah, I'd be dreading it.

I'll be looking 'who have I got to beat, who have I got to beat?' or 'They're going to beat me, don't worry about it, they're gonna beat me' or 'I've got a chance with this person'.

You're **weighing up** other people's weaknesses. It's cruel really I would be thinking 'mmm he's a bit thick, I'll beat him or he's really clever, so I won't, I won't be able to beat him'. And it's terrible to be making those sorts of value judgements about other people at nine. But you do know it. That was the culture of the school then: you told the kids at the top that they were at the top and you told the kids at the bottom that they were at the bottom.

I knew who it was ok to be beaten by at maths. It was ok to be beaten by John Smith who went on to Cambridge. It was ok, that was all right. I shouldn't be saying these names, but I can still remember them all. I can see them up there in green pen on these damn league tables. But to be beaten at maths, to be beaten at times tables by Simon Jones who I could beat hands down in any other subject. Well it was **embarrassing**.

To just not get the answer first or to just stand there—'er, er, er, er, er, er, er',—stuttering. When the question got asked and the other person's got in there and you sit down. **When you're little** like that you don't want to stand up in front of all your mates, do you? You don't **want** to stand up in a classroom full of kids because you're not as confident.

I just hated it, it was horrible. Especially when you got relegated to the second division or something. And I'm sure, the kids who liked maths enjoyed it. I don't know how the kids who weren't any good at maths at all felt about it, but I felt pretty **terrible**. It was embarrassing, yeah it was **shaming**.

That's when I really, really started to think: 'No, I'm no good at this'. It is pressure and it's scary. And it's embarrassing when you get it wrong or you get into trouble at home, you know this is when it became this whole pressurised thing. It was the pressure.

I was getting the same pressure at home about learning my tables. I would have to go home and tell me mum who had beat me. I remember one time in the bathroom, getting yelled and yelled and yelled at, for not knowing my three times table and sitting up there crying learning my three times table (*laughs*). It sounds awful, doesn't it? I'm sure my mum didn't mean to upset me like that at all, I'm sure for her she just went 'urghhh learn it' and she went down stairs and for her that was the end of it. But I got really upset about it. And it made me angry, very angry.

I'm not so much telling you my experience of maths as my experience of life and how maths fits within that. And that's really, really complicated. There are all the little bits and pieces of other things that must have happened whilst I was learning maths and whilst maths was the main thing, there were all these other things going on.

I don't remember the actual teaching of the maths; I do remember this 'being put on the spot'.

I did not have a good time with maths at senior school either. When I was about fourteen, I remember that I was taught maths by this man. We called him Speedy Brown. I don't think he had any concept of how to teach maths to children at all. He was very, very, quick. Apparently, he was this brilliant mathematician. He was a very scary, a very scary man. If you met him outside the classroom, he was lovely. But **inside** the classroom—he had the boys in tears as well.

(Intonation indicates copying the other person's style of speaking):

He would write on the board '*TO CALCULATE*' and then he'd underline it. And we would write what we had to calculate and it would be 'CALCULATION'. I only ever got it right by chance.

Sometimes we would have to copy. It would just be a complete scribble. I wouldn't have a clue what he was writing. And it would be *(raises her voice in an aggressive tone and booming voice)*:

'NOW CLASS IS THAT CLEAR AS CRYSTAL OR AS CLEAR AS MUD?'

And we'd all have to say 'crystal', *(quietens)* and it wasn't, it wasn't at all and I'd be just lost and it's—sit at home for hours with these damn things and I couldn't do them at all. It made absolutely no sense to me. Sometimes his calculations would be a page and a half long. I couldn't ask him at all.

In all my lessons it'd be 'we're going to do this' and 'we are going to do that'. Put it on the board; go through it for five minutes. And say, 'right off you go, there's twenty to do on your own', and I'd still be 'whaaat?'

I was usually bored rigid.

We worked on our own. You sat in your two, two, two, twos and got on with it on your own. I might have turned round or said to somebody 'do you know how do you do this?' I knew, I knew that copying wasn't the answer, especially seeing as the maths teacher would have known that's what I'd done. So I never sat and copied just sometimes asked 'what did you get?'

'I didn't get that. . . . try again' *(talking to oneself or thinking aloud).* Or sometimes my friends explained it to me. And it usually helped me more if my friends could explain it to me.

I was usually bored. Either I was bored or I was desperately struggling, usually it was both bored and desperately struggling *(laughs).*

I used to quite often be in tears over my maths homework.

I think the maths people enjoyed it. You see here we go again 'the maths people'. You know they are defined by their ability to do maths. And some of those kids were all the way through senior school. The thing they were known for wasn't for well you know 'because the women like him' or because 'she's good at music', but it would have been—'they're really good at maths'. They'd all be together and be sort of defined by it. That makes it a big thing, doesn't it?

I did it. I did my GCSE and got a grade C. I did the Intermediate paper[3] because I sat and revised it all from books and whatever and as far as I can remember a lot of it was really simple *(querying, confused)*.

But I still don't know my times tables. I still don't know them as I found out last night when I was trying to add up lots of long strings of single figure numbers. I was having some real trouble with that, you know I was having to count them up and doing it on my fingers.

I can do my elevens; I can do my elevens and my tens. I still have trouble with my fives *(quieter and sadder)*.

Yesterday I was adding sort of eight and four and six and I'd be like 'eight and four, eight and four, eight and four, eight and four' *(thinking aloud)* and I just think, 'I won't know'. '**Twelve**' *(with surprise)* 'twelve, eight and four—twelve and then add a six to the twelve' and then I'm thinking 'well it's a six and a twelve—six and two, two and six—eight and oh no I've lost that ten, that's gone' *(pace slows)*.

The numbers just float away, they just float away.

Even when I use a calculator I'll check three times, I **just** don't **trust** them, numbers, I **don't** trust them. They've got a mind of their own and they're just all over the place and I can't make any sense of them. At school I'd have all these numbers and I'd think I was doing a certain thing with them. And you would get an answer at the end and it'd be wrong and it'd be like 'well god, I've just spent half an hour with you lot, how can it be wrong?' They just didn't work for me.

It was a case of just randomly hoping for the best. Maths is like fog. It's waiting to trip me up. It's almost like walking through a fog and everybody else knows the way and you don't. You're just running around willy-nilly hoping you are going to get out the other side *(laughs)* and you just don't know.

Like when you've got percentages, you've got such and such over such and such times something or other. I don't know which numbers go on the top, which numbers go on the bottom, which ones you're dividing or multiplying by, which numbers? They're all just numbers. I can't apply a formula because they're all the same. I know that it might be eight and seven and six and five but to me there're all the same, they're just numbers. I need definite things that my mind can hold on to.

I've been fobbed off. When I asked 'why', they said—'just because it is.'

(In the next part the teacher voice begins impatiently, the learner is calm as the dialogue continues the learner's voice becomes more impatient.)

'But why?'

'Don't ask questions like why, it just is, Pythagoras's theorem just says it'.

'Well why?'

'Just because it does, that is what it adds up to'.

'Why?'

'Because it does'.

'Why?'

A lot of maths is 'because it is'.

'Well why is pi three point what ever it is?'

'Because it is'.

'But why isn't it fifty six?'

'Because it isn't'.

'Why?'

What I mean is, it comes back to '**why?**'

I have trouble with the fact that that they can be right and wrong when a man made numbers. A person decided that pi is this, that and the other. How can he know really? Because he made it up didn't he? Made all the laws as well and I know that they all fit together and I find it hard to get my head round how somebody can discover a new mathematical law.

And it's just because it is.

And it's us that decided we needed numbers in the first place. Well we do. I need numbers too. I need numbers to think how much did I get paid for this week's work. You know I'm not saying I don't need numbers, I do. But if it's a man made concept then why can't we say 'well because of this?'

I hate to say it—it's this sort of maths male world thing. Yes, maths—it's all been male dominated. I don't think I've ever had a female maths teacher. Even my female teacher at junior school spent all her time teaching maths to the boys. I think there was only one of my friends who was a girl who was very, very good at maths, all the rest were boys. I don't know, it's very much, such a straightforward, cold, impersonal kind of thing. That's the thing that bothers me with maths, it's not more appropriate, less appropriate, different—it's right and wrong, black and white.

That's another reason I had trouble relating to it. Maths is straight lines and I think in circles.

I'm very good at creative stuff. In English and that sort thing, there are feelings, emotions, trains of thought, frames of reference. It's all circles, there isn't necessarily one that has to be done before the other. It can be done non-chronologically, as it comes to mind. You know, you're piecing something together from all these little things and it doesn't really matter sometimes which one comes first. Whereas maths, I think you've got a first bit and then a second bit you have to do and it's a straight line. You need a tidy mind *(moves her hands forward in straight line, palms facing outwards, pushing away from the body)*.

I suppose this is it, if I've got a collection of numbers, I can't get them in a straight line, they just go round in circles *(laughs)*.

I am right at the bottom when it comes to maths *(laughs)*. And it is not as important to who I am now. The fact that I can't do maths is something everybody else has learnt to live with about me.

But I've had jobs before where there is no till and you have to add up in your head. And I have been able to do that probably because I haven't worried about it because if you get it wrong it doesn't really matter *(laughs)*.

You know I think I've blamed the numbers and blamed the maths and in fact I couldn't do it and that's stayed with me. When in fact it's not the numbers at all. That's the way people behave towards their computers isn't it, 'oh it's just crashed for no reason and I've pressed this and it's done that'—but you've done something. So you can only react in a way to what you've done Because you're the bad workman. And I feel that all along I've blamed the tools and it such a part of it now, but I realise now on an intellectual level if you like, it's the bad workman, it was always the bad workman. And that was me. You know and possibly the people who tried to teach me but I'm well aware that it's not the maths that's that the problem. But as a kid it would have been those horrible numbers, but it's not, I know it's not that.

Totally nine again, that is what I am when it comes to anything to do with maths. That's how old I am again. Worried about it, yeah, it feels like you are like some big kid who still can't do something.

(Boylan, 2004, pp. 160–166)

COMPULSION AND STRUGGLE

A sense of compulsion pervades this narrative. Louise repeatedly tells us she **has to** do the practices that would most naturally be seen as mathematical, such as counting, 'calculation', using formulas, percentages, and so on. Even more forcefully, she tells us that there is no choice about engaging in the pedagogical practices that, although not essentially mathematical, are intrinsic to her mathematical experience, such as being put on the spot, copying, working individually, and so on.

Louise's relationship to mathematics is threaded with images of conflict and struggle. Louise struggles with the mathematics as given and in the development of her own mathematics in relation to it. She identifies herself as struggling for meaning. She articulates her own inquiry into the nature of mathematics, addressing a central issue in the philosophy of mathematics. She stories this struggle as a dialogue between herself and what we take to be a fictionalised teacher who stands for her experience of teachers and school mathematics. This teacher 'fobs her off', will not or cannot respond to her requests for explanation—explanation that might support her need to make sense of her mathematical experience, of why mathematics is what it is.

It is in this context of compulsion, lack of choice, and struggle that Louise carries out her identity work and stories herself. We choose to highlight two aspects of the constraints upon this identity work.

MEASURING

First, Louise reveals for us aspects of the process by which she internalises the school mathematics practices of being ranked. These are practices which exist whether she wills it or not, practices of which she cannot but be part. In the primary school league table, she is in division one, but she is 'usually hanging around at the bottom'. She is not one of the 'maths pundits', and there will never be any point teaching her long division. We might hear this part of her tale as simply a very unfortunate encounter with a teacher with a particularly heightened sense of labelling her pupils, of telling 'the kids at the top they are at the top and the kids at the bottom that they are at the bottom'.

However, we contend that it highlights and focuses attention on an aspect of school mathematics practices that requires more attention: the experience of being measured and of measuring. We use the term 'measuring' to highlight an aspect of school mathematics identity work that is pervasive and personal. We might commonly suppose that the actors in the mathematics classroom who do the measuring are most obviously the teachers: after all, they set the tests and appear to be the authors of the scripts/stories and thus the roles that learners are required to play and internalise—'division one', 'bottom set', 'no good at maths', and so on. However, learners are not simply measured, but are actively engaged in measuring themselves in relation to others: 'weighing up' others' relative strengths and weaknesses. In addition, they are themselves objectified as measures through the process of ranking. To participate in school mathematics is not only to be measured, but also to be in some sense a measuring instrument.

Measuring not only happens in relation to other people, it also happens in relation to measuring oneself against both the social practices of mathematics and the objects of mathematics themselves. One is being measured and thus measuring oneself by one's ability to participate in the league table competition, use formulas, and so on. Louise is fearful as to whether the numbers can be beaten and whether she can avoid being tricked or tripped up by them.

The striking ways that Louise speaks about mathematical objects is an aspect of the second feature we highlight about the possibilities for her identity work in relation to mathematics.

BEING WITH THE 'OTHERS' OF MATHEMATICS

Louise's story reveals a wider 'truth' that, in our life worlds, our relationship with objects, whether physical, mental, or social, is multifaceted. Her account of her life world certainly adds weight to the longstanding contention that the alienation of some learners of mathematics arises at least in part from the gap between the publicly sanctioned discourse of logic and rationality and the emotional experience of mathematics (e.g., Boaler & Greeno, 2000; Walkerdine, 1988). However, we believe that those parts of her account where she most directly addresses her relationship with mathematics and mathematical objects and practice as *she engages with them* offer an insight into an aspect of the experience of mathematics that has not so far featured prominently in the literature.

For Louise, numbers and other mathematical objects are different from the Platonic ideals, reifications of social practices, or other abstract ontological constructs that are often proposed in academic accounts of the nature of mathematics. Rather, her relationship to them at times is, in phenomenological terms, akin to 'being with others' (Heidegger, 2000/1926). These are others who have 'got a mind of their own'. It follows from this that, for Louise, it is not simply that there is an emotional or affective dimension to mathematics. Rather, just as emotionality is intrinsic to interacting with, and relating to, other people, so it is intrinsic and inseparable from her experience of mathematics. In Louise's world, these emotional qualities are part of the struggle, with all the visceral experience of tight stomach, dry mouth, and tensed muscles. Louise does not have a choice about whether or not mathematics has this emotional quality. Thus, the story told to her about mathematics in school is one that necessarily alienates and disconnects her from her experience and reinforces interpersonal, intrapersonal, and existential isolation (Yalom, 1980).

We have proposed that Louise's story highlights an important aspect of the mathematical experience: the way in which mathematics and mathematical objects are experienced can be similar to the way we experience other people. We do not make the claim that this is true for all people. Perhaps it is an aspect of being someone who, as she tells us, is 'very good at creative stuff'.

However, we note that the ways in which professional mathematicians describe their own experience of mathematics, although much more positive, share something of this quality of an intimate relationship with mathematics (Burton, 2004a). Personally, as two relatively successful learners of mathematics, we reflect on our own engagement with mathematics and find that mathematics presents itself to us in a way that is intimate rather than abstract. We are fortunate that our mathematical 'relations' are mostly experienced as allies with whom we can usually choose how we interact. This is not the case for Louise. Numbers in her life world are sometimes uncooperative, uncontrollable, and untrustworthy *others* who deceive and trick. She is subject to them and to their capriciousness.

FINAL WORD

So, finally, what of Louise and choice and mathematics? In the layered process of creating stories about Louise from the stories she told about herself, we found ourselves drawn towards writing a salvation narrative, pointing up the choices she is able to make as she retells her experiences. It is the case that she does not only fit into the space created for her by others; for example, she creates some spaces for anger and for personal authority, for questioning and for redefining. But Louise's story is one of pain and alienation, and it is in this unchosen context that she does her identity work. The fact that such identity work happens means that she does, in spite of her experience of compulsion and struggle, have scope for choosing. However, we cannot ignore and she cannot escape her final damning judgement on herself as 'the bad workman'.

NOTES

1. All names in the research artefact are pseudonyms.
2. All initial teacher education students in England and Wales are required to pass 'numeracy' tests before they can qualify to teach (Povey, Boylan, Elliott, & Stephenson, 2000).
3. In UK GCSE mathematics examinations, taken at the end of compulsory secondary school, students were entered for one of three levels of examination, Intermediate being the middle level.

6 Choice
Parents, Teachers, Children, and Ability Grouping in Mathematics

Peter Winbourne

INTRODUCTION

This chapter draws on research in a school in which parents are very much involved in all aspects of the school's strategy and policy. When the school was established, mixed-ability grouping was central to its organisation. Recently, the school has moved to a system where Year 7 has been streamed: children now work in groups that are seen as fairly homogeneous in terms of ability. The grouping system is described as follows: one 'Accelerated' group, two 'Accelerating' groups, four 'Standard' groups, and one 'Nurture' group. The main criterion for grouping students was average level of attainment across subjects of the National Curriculum[1]. The school used other data to group the students, including Cognitive Ability Tests (CATs), Reading Test Scores, and internal assessment. It was introduced in April 2008 for Year 7 (Y7) students and viewed as a pilot until the end of the summer term.

The choice to move from mixed ability to streaming was taken by senior managers at the school, who stressed it had been constrained by real and practical considerations over which the school had no control. This decision was based on their observation and monitoring, consultation with teachers, and perception of the views of parents. The school's observations and monitoring had indicated that the teachers were teaching to the 'middle' of the groups. Teachers had expressed concerns about the quality of children's experience at the school; they also found the spread of abilities too wide to cater for in mixed-ability groups. There was a belief within the school that lower achievers needed to be caught as soon as possible if they were to be afforded proper access to the curriculum throughout their time at the school. Parents had made no direct request for this change in grouping, or to opt for the particular choice of streaming, but this school, like most schools in England, is engaged in a struggle within a discourse of performativity (S. J. Ball, 2008), and there was a clear perception of pressure—from parents, from Office for Standards in Education, and from government: talking about curriculum, the head teacher said, 'every school has a lot of pressure on it. 12 of my 20 governors are parents of children in the school. That means you get a lot of pressure on you'.

There are powerful discourses at work here which serve not only to frame discussion, but also to shape the identities of the people involved. The naming of these groups is an indicator of one of these, suggesting a language for describing education framed in terms of pace and hierarchy, rules that Bernstein (2004) argues apply in any pedagogic context. Indeed, in their discussion with me, most parents, teachers, and children either could or did not speak of each other without evoking some image of speeding through the business of schooling.

ETHICS, LIMITATIONS, AND FOCUS

I should be clear about what the focus of this chapter is and what it is not. The focus is not the school or what might be seen as its particular and distinctive features, or the process of choosing to move from mixed ability to streaming for Y7, which, from here on, I will call the 'choice'. The school does, indeed, have distinctive features which could be seen as explaining what central actors there have seen as a need to make choices; these features also, I think, explain how it has been in a position both to make, and be seen to make, the 'choice'. That is, parents, teachers, outside observers (people like me included), advisers, and children, knowing what they do about the school and its context, would all have been in a position to understand the 'choice' and judge its legitimacy in ways that cannot be rendered here. These actors could and would identify real constraints upon the school, in terms of resources of all kinds, which have meant that the scope for choice has necessarily been limited; indeed, they have referred to these constraints in their conversations with me. Writing about the nature of these constraints and opportunities would be an interesting activity, but I have agreed with colleagues and all taking part that the school should not be identified in my writing, and anything more than the brief details I have given would identify the school. As the head teacher herself said, 'It's incredibly interesting . . . there is something so interesting to be written about all of this . . .', and so there is. I hope, of course, that this chapter will be of interest, but, at the risk of frustrating the reader, it is not, and cannot be, about the particular set of circumstances to which the head teacher was referring here.

Whatever the circumstances of the school, here was a set of people experiencing changes in their lives that have resulted from the 'choice'. In this chapter, I focus on some of the Y7 children and what impact the 'choice' appears to be having on their lives: how they experience school, how they see themselves as learners (and learners of mathematics), and how they are seen by others. I seek to contextualise their experiences through the parallel, though less detailed, exploration of the experiences of some teachers and parents. This focus may afford some insights and some small glimpses into the actual process of the 'choice': when people tell their stories, they include detail of their responses to 'the process', in what way they were involved, maybe; how their voice may or may not have been heard; and how the process of choice itself may have contributed to the development of some aspect of their sense of self, their identity. The 'choice' is present, in some sense, as a constraint that has shaped both people's experiences (of school, of education, and of themselves in these contexts) and as a lens that has shaped my perspective and that of interviewees as we have probed these experiences. For those familiar with her work, it may be helpful to think of the sculpture of Rachel Whiteread: delineating the space inside a room focuses attention on that space; it is a space that could not exist without the constraints of the room that acts as boundary and so forms it, but the room is not the object of attention; the room is there to be reconstructed, glimpsed; but we are not invited to do this, but rather to attend, in ways otherwise unavailable, to the space.

Of course, the same ethical considerations constrain my selection of data from the mass of interview transcripts. This leads me, inevitably, to selection and editorialising, but then such selection and editorialising is ever present in any kind of study, and in particular in any sociocultural study: You can try to capture lived experience, and, if you're lucky, you will render some so it can be glimpsed, but there's always choice, always editing. This means that the fact of my 'editorialising' needs very much to be kept in mind as this chapter is read. I talked about my editorialising with all participants as we worked together to make our texts: I would have to make choices, in some cases because of my own agenda, in others because of my perception of the agreed ethical constraints upon the research activity. I will explore 'editorialising' further in the section on methodology.

THEORETICAL PERSPECTIVE

Mathematics has figured prominently in much of the research literature on ability grouping. Of the departments at the school, the mathematics department found mixed ability most problematic, and, as might be expected, mathematics figured prominently in the accounts of those involved. Here I use these accounts to map in some detail the part that mathematics (or conceptions of mathematics and its importance) in particular and wider experiences of schooling, mediated by a range of artefacts, has played both in experiences of the 'choice' and, more generally, in the lives of some of the parents, teachers, and children involved.

Here I am using notions of identity worked out in recent writing, but also influenced by Gee (2000–2001):

> Being recognized as a certain 'kind of person', in a given context, is what [Gee] mean[s . . .] by 'identity'. In this sense of the term, all people have multiple identities connected not to their 'internal states' but to their performances in society. (p. 99)

I continue to make use of a view of 'identity' as the aggregation of the smaller 'becomings' (or identities) identified with a learner's participation in a multiplicity of communities of practice, local and not so local, some of which are locatable within school classrooms and most not (Lave & Wenger, 1991; Winbourne & Watson, 1998a). Holland, Lachicotte, Skinner, and Cain (1998) point out that 'Identities become important outcomes of participation in communities of practice in ways analogous to our notion that identities are formed in the process of participating in activities organised by figured worlds' (p. 57), and I make use of figured worlds, too, where I think this is helpful.

Community of practice is a problematic notion. I have written about this at length elsewhere (Winbourne, 2008; Winbourne & Watson, 1998a). I will not revisit this here, save to acknowledge the recent critique of Kanes and Lerman (2008). They are critical of situated accounts that make use of concepts of community of practice, but which lack

> a sense of the institutions around both the teacher and student in contexts necessarily broader, but containing the classroom experience itself, and a sense of how these maintain and regulate the life of classroom practice. (p. 309)

Here I set out to provide a sense of the communities of practice and the broader institution that both (a) contextualise the activity of children, their parents, and their teachers; and (b) constitute and are constituted by aspects of their developing identities. This may be at some cost, however, as there is also a perhaps necessary loss of focus on the learning and teaching of mathematics.

EXPERIENCES OF ABILITY GROUPING

This section serves partly as a brief, selective review of some of the literature on ability grouping, but mainly as the start of the presentation of the study. I talked with two parents[2], three teachers[3], one deputy head teacher, the head teacher, and five children[4]. Some of these people wanted to talk about evidence for and against ability grouping in their conversations with me, in some cases, quite reasonably, wanting to know where I stood, as in these extracts from interviews with parents Martine, whose son is in an 'accelerated' group, and Clare, whose daughter is an 'accelerating' group:

Martine: What do you think, as an adult person as to ability groups?

Peter: I think that these things are a product of particular ways of thinking about schools and teaching and learning; they're kind of an inevitable product of ways that we understand schools. . . . There's quite a lot of work that's been done looking at the effects of various types of grouping and there's not a lot of evidence in favour of ability grouping really. There's some; there's a lot of research done, but the evidence is not clear-cut[5].

Clare: Of course, what I'd like is systems that work for every single child in the school; that is what we should have. As a governor that is what I would say. But I am also concerned that it works for [my daughter]. So, from thinking that streaming was the answer, maybe it needs to be all three things[6], focussed on the abilities of each child, which is very difficult . . .with all the teachers completely rushed off their feet all the time . . . Is there really not much evidence that any of these things make a difference then?

Peter: Well, quite a large quantitative study by people at the Institute[7] used quite a large survey looking at the amount of setting that children experience up to GCSE and relating that to GCSE success and they find no obvious patterns, and that's a large quantitative study. Other qualitative studies[8] look at children's experiences and stories such as [your daughter's] are not uncommon, where she talks about her own experiences.

[Earlier in her interview, Clare had said: last week I actually went to see the head about a number of things, which included talking about how the children were being taught. The night before I went, I talked with [my daughter], it was kind of the first time I had talked to [her] about it really, and she was actually saying that she actually preferred it the way it was before. She said that her and two other girls are like the top of the class, which is quite unusual, because [my daughter] and these two girls aren't actually that bright, really. What she seemed to be saying was, in essence, that it wasn't very motivating because there wasn't any kids above them. And she said when they were in the mixed ability groups—this is obviously a 12 year-old—but, when they were in the mixed ability groups she felt more inclined to help the ones that were struggling and she doesn't now.]

Peter (continues): My own view is that it's very much a product of particular ways of looking at school. If we have to look at school, in terms, perhaps quite rightly, of results, then this may seem like what you have to do. But there's also evidence . . . that if you look at school in other ways[9], like in terms of democratic participation, then the results also come. . .

> Clare had used the internet to read up on some of the evidence herself. She had used the Department for Children, Families and Schools (DCSF) site and searched under streaming:
>
> Clare: There was a research study on there, but there wasn't much there ... and that study said what you've said, that streaming wasn't necessarily indicative of good results, and it was all a bit inconclusive actually.

I am not sure what research Clare had observed. However, materials prepared for Ruth Kelly, when she was Secretary of State for Education, and which were used as background for a speech she gave in 2006, show the extent to which grouping by 'ability' is central to official discourse, in spite of the lack—well known to her department—of supporting research evidence:

> Ability grouping—for example by setting or streaming—can be used to help tailor learning. Many schools use such grouping, although analysis does not find a significant difference in average attainment between setting and mixed ability teaching. There is some evidence that setting can widen the attainment distribution. This may be the result of poor teaching in lower attainment groups and pupil motivation.
>
> The challenge is to ensure that setting, as part of personalised learning, is implemented in the right way to tailor learning effectively and not in a way that will widen the attainment distribution.
>
> Departmental analysis currently underway is exploring methods of ensuring that setting can benefit more disadvantaged. (DCSF, 2006, p. 64)

Kelly and her advisers were either unable or unwilling to think about education in terms that do not include ability grouping. For similar reasons, her successors may not be able to hear the voices of those engaged in the current, large-scale Primary Review who call for much more effort 'to be directed to the development of classroom-based social pedagogy, including the effective use of pupil grouping' (Blatchford, Hallam, Ireson, Kutnick, & Creech, 2008, p. 31).

METHODOLOGY

I set out to probe how children, teachers, and parents had experienced the 'choice'. I wanted to enable the voices of these participants to be heard, aiming to involve children, teachers, and parents in aspects of the research and, to the extent that this was possible over such a short period of time (2 weeks), as coresearchers.

I wanted to see how what could be thought and said about schools is framed and constrained by current educational discourse. I wanted to get some insights into some of the possibilities for being and becoming, glimpses of the formation of discourse identities (Gee, 2000–2001), that might be afforded by talking with people about the changes in practice that have attended the 'choice'.

I felt that some insights of this kind might be possible through inviting interviewees—teachers and parents mainly, but children to the extent that this might be possible—to entertain the possibility of the world of education being other than it is; in particular, how might this world look were we to believe that, rather than regroup children into 'ability' groups, we should establish groups that were as diverse and heterogeneous as possible.

I wanted to encourage open and honest participation, and I was concerned that children, in particular, should not feel inhibited when it came to talking about their experiences of the change to ability grouping. My experience and that of fellow researchers (S. Weller, personal communication, 7 May 2008) has been that, in such circumstances, children can feel unable to talk, even anxious. So, at the start of the first interview, rather than ask children directly about their experiences, I took an indirect—though, to the children, quickly transparent—approach and used a vignette to invite them to speculate on the feelings of others in a similar situation:

> *Peter:* I am interested in the experiences of children at schools where the ways that they are grouped for their lessons has been changed. At one school—I won't say which—I know that some girls and boys found that they were moved from a group where they worked together with children of all different levels of achievement and abilities to a group where they were working with children who had all achieved roughly the same levels. How do you think they might feel about that?

Processes of text negotiation and editorialising were central to my way of working and were made explicit to interviewees from the start. I used a process of text negotiation, similar to those I have used with colleagues in recent research (Wilson, Winbourne, & Tomlin, 2008) and about which I have written more explicitly elsewhere (Winbourne, 2007). My aim was to enable participants to feel that their voices would be heard legitimately in this text and to secure greater validity. All the teachers and children whom I spoke to said that they trusted me to choose which of their texts to include. Parents and teachers were keen to read the chapter; the children wanted to see the book[10].

The following exchange, during the second interview with the five children, exemplifies the text negotiation process:

> *Peter:* You see, . . . when I was talking to Eugene, Chris and Aaliyah, I asked this question, 'Have people in this school had problems?' and Chris said, 'Yes, people in the lower group'[11]. If, in what I

write, I just put that, what are people going to think, if that's the only bit I wrote about in my chapter?
Eugene: That it's a bit boring . . .
Chris: That I'm in it. That I am in the lower group, cos I'm not.
Eugene: But that's why we changed our names, cos they don't know that you're Chris.
Peter: That's entirely my point: if that's what I write, Chris said that they might think that he was in the lower group . . .
Chris: I didn't just say that, I also said that sometimes people are kind of bad, yeah, but when they're in their class they're proper smart . . . their friends might think they're neeky[12] cos they're smart in class.

The process was intended to support a hermeneutic cycling towards text that would represent an acceptable truth for participants (Van Manen, 1990). In the example I have just given, as the children discuss the text, they speak of each other in terms of how others would see and speak about them; they reveal some aspects of their sense of who they are: within some of the practices that intersect at the school and the figured worlds which are the spaces for their development of self. In the following, I present more texts that I have assembled through my coding[13] process, aiming to represent aspects of children's experiences of the 'choice', looking for further signs of how they see themselves and how their identities are produced in practice and in discourse. I use extracts from interviews with teachers and parents to provide a sense of the broader institution and community. In my selection, I have been guided by the need to show the position that mathematics might have in all of this.

TEXTS AND CONTEXTS

Being standard, accelerating, and accelerated

Aaliyah's world is one where being seen as high achieving is important; Eugene makes clear that it is important for him, too:

Aaliyah ('accelerating' group): I am not the highest, I am the one underneath it, but I am at the top of my . . . I am going to move up into the highest one here . . .
Eugene ('standard group'): Yeah, I want to move up into another class. . . .
Aaliyah: . . . I am going into the highest, yeah, all the teachers are telling me that I am on my way.

The 'choice' may not have changed Aaliyah's ambition or view of herself, but it has afforded her new ways of talking about herself, clearly framed in terms of hierarchy and with a sense of movement and pace. Her omission of 'in' ('I am not the highest') may be due to her way of speaking, but it is striking in this context. Kwame ('accelerated' group) makes similar omissions and very clear reference to pace and speed:

> *Peter:* When there was that broader mix in class, were people expected to help each other?
> *Hera ('accelerated' group):* Yeah.
> *Kwame:* Yes. But sometimes it slowed down the learning. If, let's say in maths there's something called accelerating, accelerated and standard, yeah, and if some people are standard, accelerating and accelerated, and the teacher would have to explain all of them, so we wouldn't get that much work, but, if, now, you're in the highest group for maths, accelerated, they don't have to go through all the accelerating, accelerated and standard a same thing (for all of these groups). It makes more time to learn . . .

Kwame used the names of the groups as identifiers for himself and other children; he seemed to think that these names applied only to mathematics. A comment of the Deputy Head teacher[14] with whom I worked gives some sense of how these are used in the school and, at the same time, suggests both the centrality of mathematics in the children's experiences of the 'choice' and, possibly, a parity of level and speed:

> I think only Maths do use the names—we deliberately downplayed them in an effort to avoid labelling—but obviously the children are fully aware about which group they are in; although as I said previously I am interested in increasing the use of those terms to make the children aware of the level of work they are doing across the curriculum—so it is consistent. Brett [Head of Mathematics] picked up on it because it was what he was familiar with at [his previous school[15]].

Being Gifted and Talented

Probing to see if children in the school were generally happier now—after the 'choice'—elicited responses like this:

> *Chris ('standard' group):* We would always have jokes with teachers. Now we're an ordinary class.
> *Aaliyah:* My class is totally different. More people . . . I feel proud of myself because I am in the second to highest group. We all kind of know that we are gifted and talented and can produce excellent work.

We always work together and it's always smiles. It's not like we are hanging our heads down because we can't achieve anything.

Being identified as 'gifted and talented' is important to Aaliyah, but the new grouping highlights some confusion. Recently, a trip to Sussex had been organised for children identified as gifted and talented:

Eugene: I would like to go to Sussex.
Peter: Do you know who the children are who are Gifted and Talented and on that trip?
Eugene: I know most of the people who are going to Sussex.
Peter: Which groups are they now in?
Eugene: Some of them's in my class . . .
Aaliyah: . . . most of them are in the top set and my set
Eugene: . . . some of them, two are in my class, I know, yeah, cos [boy] was going to Wales, but he didn't want to go Wales, so [other boy] is going in my class.
Peter: So, he's in your class [i.e. Eugene's] and he's gifted and talented; how does that work?
Aaliyah: I don't understand that bit. There must be gifted and talenteds in all different sets; but obviously he's a gifted and talented for his level; but maybe he's a gifted and talented for his level, but he can't face the gifted and talented at my level, I don't know, I don't know.

Kwame's Friend: the all-rounder

I didn't meet Kwame's friend, Darrell[16], but he 'turned up' first in conversation with the children and was subsequently 'identified' for me by Brett, the head of mathematics. He appeared, I think, both as Kwame's true friend and as a character in the children's figured world; his combination of respect on the street and recognised academic 'ability'—enhanced as a result of the 'choice'—captures, I think, some of the essence of that world.

In the first interview, Kwame said:

I don't miss all the people in my old class. . .just that people do miss their old friends from their old class. If my friend (Darrell).. if my friend was still in my class, yeah, . . .

Later in the same interview:

Kwame: Sometimes, yeah, in my (maths) group, yeah, there was five people went to a different classroom, like to do higher stuff sometimes, to do extra maths. Four of them . . . are in my class . . . but, my friend, Darrell, I think he got like, I don't know, I

68 *Peter Winbourne*

> think he got 4–4-5[17] or 5-4-5 , as well, then he improved, but I don't know why he's not in my class.

Darrell 'turned up' again on the morning of the children's second interview:

> *Chris:* Some people didn't expect some people to be in the like the group that they're in, like the higher group that they're in . . .
> *Kwame:* Like some person, loads of people thought he wasn't that smart, cos how he acts outside of classes, ..people think that he's not all that smart . . .
> *Aaliyah:* I think we're all talking about the same person..
> *Peter:* How, why are they changing their view of him now..?
> *Chris:* Cos people see what class he's in . . .
> *Aaliyah:* He can get even more respect now . . . that's showing that he's like an all-rounder..

Later that morning, Brett was reading some of my notes from our first interview. I had drawn his attention to what he had said about the 'few children (seven) for whom the change had really not worked and mixed ability would be better, but most are happier and feel like they're learning which is important. Those seven were happy and were working in a pocket of children including brighter children and were being brought along quite well':

> *Brett (reading on from my notes):* 'One was the brightest, and is now in a competitive environment in his new group. It turns out to be Kwame's best friend, who is now in second group (and has said he missed.—see K's text)'.

Brett also reads the comment I have added where I draw attention to what Kwame had said about Darrell and asks, 'What's that?':

> *Peter:* Oh, about, cos one of the things that Kwame said was that one of the brightest boys, in fact it's Darrell, is in this different group . . .
> *Brett:* Initially he was going to be in one of the standard groups—this is why I feel like we've got the groups really right actually— Initially he was going to be in one of the standard groups and he's got the highest CATs and I thought he was going to die in there. I used quite strong language, like that, because I think it is like killing a kid.

WHAT ELSE COULD SCHOOL LOOK LIKE?

The world of the school is, indeed, such that the 'choice' may better be seen as an inevitable product of the powerful discourses to which I referred earlier. These discourses extend to the children's homes, where they also powerfully shape identities (Hughes & Greenhough, 2008). In some sense, children 'bring' these identities with them to school (Winbourne, 2008). Discussion with Martine provided some insight into the constraints upon such shaping of identity:

> *Peter:* To think out of the box a bit, imagine, perhaps . . . what might be different about [school] if, rather than thinking, oh yes, . . . we must put people of the same ability together, we were to think the opposite: which is to say we **must** have a range of abilities in each group?
> *Martine:* Well, what level do you teach at?
> *Peter:* I was just thinking what the school would look like if that's how you felt you needed to respond to it, do you know what I mean? . . . what would school be about?
> *Martine:* Oh, I don't know .. If children have the ability, it makes no difference . . .
> *Peter:* Yes, I suppose, what I was saying is, if it were the reverse, you know, somehow because of the way school would be in this different world, as it were, we would feel, yes, we have to mix the groups up . . .
> *Martine:* I think yes perhaps . . . What, are you talking in ability?

In our conversation, I found that Martine and I could not talk of school as other than structured through 'ability'. Brett and I were able to do so. In response to similar questioning, Brett offered a vision of school:

> that dealt with first, maybe, learning how to learn—with actual learning that was measurable secondary to that, still assessable, but later on. Not speeding through stages of development, setting up structures. Involving children in how to learn.

This is a vision strongly supported by Boaler's research (Boaler, 2008; Boaler & Staples, 2008). It is also a vision which Brett shares with his colleagues and which, in their conversations with me, is seen as out of reach. Were it to have been realised in these children's experiences, they might have spoken very differently about themselves and appeared differently in these pages.

ACKNOWLEDGMENTS

I would like to thank John Wilkinson who has collaborated closely with me on the research for this chapter. Thanks, too, to Yvette Solomon, a supportive editor and critical friend.

NOTES

1. 'Accelerated' group—average Level 5b; 'Accelerating' groups—average Level 4b; 'Standard' groups—average Level 3b; 'Nurture' group—Ns or Bs in English.
2. I spoke with parents by phone; I did not interview their children. Interviews with staff were one to one. For the first interviews with the children, I spoke with Aaliyah, Chris, and Eugene together, and then with Hera and Kwame together; for the second interviews, I spoke with the five children together.
3. Brett, Head of Mathematics; Juliet, Head of English; Sophie, teacher of PE and mathematics.
4. The children and the teachers decided the names by which they would be known in the text. I chose the parents' names.
5. I was thinking here of reviews such as those of Department for Education and Skills (2005), Ireson and Hallam (1999) and Sukhnandan and Lee (1988).
6. Clare means streaming, setting, and mixed ability.
7. Ireson, Hallam, and Hurley (2005)
8. I was thinking of Boaler, Wiliam, and Brown (2000).
9. Here I was thinking of Boaler (2008). Had I read their report then, I would also have thought in terms of social pedagogy and Blatchford et al. (2008).
10. They were astonished to hear how expensive books like this can be.
11. At this time, we were all looking at the transcripts of the first interviews from 2 weeks earlier. I had highlighted some of the text in yellow, adding comments to show how I might use it. I had done this precisely to encourage further discussion of voice, legitimacy, and participation at the second interview. The children had read these and discussed them during the first part of the second interview.
12. Neek: 'A cross between a nerd and a geek' (Urban Dictionary, http://www.urbandictionary.com, accessed 2 February 2009)
13. I transcribed the recordings of interviews and coded them using TAMSAnalyzer (http://tamsys.sourceforge.net/) and open coding.
14. A close collaborator in this chapter, who is well aware of my 'agenda'.
15. A very successful comprehensive school in the same Local Authority.
16. Name was changed by me.
17. These three numbers are the National Curriculum levels that Kwame thinks Darrell reached in the tests that they all took at the end of their Primary schooling.

7 Special Cases
Neoliberalism, Choice, and Mathematics

Heather Mendick, Marie-Pierre Moreau, and Debbie Epstein

We live in 'new times' defined by a neoliberal political framework (Hall, 1991). This has led to market economics expanding into public sector areas including education (S. J. Ball, 2007). Consumer choice is central to this; it makes the free market function. Choices are not simply acts of consumption; they are also a means of making one's self. 'Individuals are to become, as it were, entrepreneurs of themselves, shaping their own lives through the choices they make among the forms of life available to them' (Rose, 1999, p. 230). Within neoliberalism, we all make choices in order to choose 'who we want to be', to regulate and govern ourselves in an era of apparent freedom. We are confronted by an array of possible choices, all of which we are asked to make. In education, we see this from choices of schools to choices of subjects. In this chapter, we ask: What does it mean for students to make subject choices within this framework of compulsory choice and entrepreneurship? How do students negotiate choices to continue with mathematics or not within this context?

The data and analysis we draw on come from research exploring the ways that learners construct their identities in relation to popular cultural representations of mathematics and mathematicians. This research combines data from a quantitative survey, qualitative focus groups, individual interviews, and popular texts. In this chapter, we draw mainly on the individual interviews as issues of choice were central to these. We interviewed twenty-six 15- and 16-year-old school students across three mixed comprehensives in England and 23 university students in England and Wales (11 final-year mathematics undergraduates, 9 undergraduates, and 3 postgraduates studying humanities and social sciences). England and Wales have an increasingly stratified higher education system (Archer, Hutchings, & Ross, 2003) so we chose students from contrasting institutions: 12 participants were in Russell group, 'elite', research-focused universities and 11 were in post-1992, 'mass', teaching-focused ones.

We asked interview participants about:

- the place of mathematics 100 years into the future and in a hypothetical world where mathematicians appear regularly on television;
- their relationships with mathematics, any changes in these and any associated memories;

- the reasons for their educational and career choices;
- how to arrange a series of images of people and mathematics in order of likeability and a series of images of mathematical artefacts in order of their 'maths-ness' (you can view these at www.londonmet.ac.uk/mathsimages);
- whether everyone can do mathematics or whether class, gender, and 'race' make a difference.

The interviews, which lasted from about 25 minutes to over 1 hour in length, were audiorecorded and transcribed. Participants chose their own pseudonyms.

We analysed the interviews thematically, dividing the two groups of participants into those who had chosen or were planning to choose mathematics and those who had/were not. Within each group, we looked at the patterns of responses and discourses. We then made connections across the themes. This chapter is an example of one such set of connections. In the first section, we outline our discursive approach to choice and identity. In the second section, we look at the reasons the school and university students gave for all of their subject choices, and we show the centrality of enjoyment, 'ability', and identity to these. We argue that the centrality of these three factors to people's choices relates to particular understandings of the self and, specifically, to whether, through choice, one can realise the neoliberal imperative to 'be somebody'. In the next section, we look specifically at mathematics and ask in what ways such selves can choose this subject. We do this by focusing on the 11 interviews with mathematics undergraduates. We show that their choice of mathematics relies on the way that a position as mathematically able confers an identity as different and special. This has consequences for mathematics and society: it excludes many people from mathematics and disproportionately excludes particular groups. It does this because difference/specialness relies on discourses that not everyone can do mathematics, and we show how these are held in place by boundary keeping practices around mathematics.

THEORY

There is a vast array of discursive approaches (MacLure, 2003; Wetherell, 2001). Ours draws on work by Michel Foucault, in which discourses are stories about the world. But they are more than this because we think of stories as something that we make up, whereas discourses are collections of practices that make up us and our world; they are 'practices that systematically form the objects of which they speak' (Foucault, 1972, p. 49). Analytically, this means that we do not ask 'Is x true?', but rather 'What makes x possible?' and 'What are its effects?' For example, we do not ask 'Is it true that some people have greater natural ability in mathematics than others?', but rather 'What makes it possible to think about some people having greater natural ability in mathematics than others?' and 'What effects does this idea have' on the ways we think of mathematics, ourselves, others, and the relations among these? Similarly, reformulating our analytic questions around choice, we must move from, 'Why did someone choose (not to) continue studying mathematics?' to 'What makes it possible for us to think of someone as making a choice (not) to do mathematics?' and 'What effects does this have?'

This is a switch of focus that foregrounds not the individual and their 'choices' and 'abilities', but the ways that people are assembled as (not) choosing mathematics and as un/able through patterns of relationships, materiality, and so on. Artist Rachel Whiteread's work provides an analogy. She represents objects—rooms, staircases, and tables—by making sculptures of the space around them[1]. It is remarkable how different something looks and feels when represented in this way. So we need to explore people's choices as much by looking at the space around them as by looking directly at the space they occupy. This shift in focus is important because:

> If we emphasise relations over subjects, the solidity of subject and object boundaries begins to fragment. I am interested in that fragmentation because it allows us to focus on the complexity of what happens in ways that show us connection rather than separation into discrete persons acting on objects. (Walkerdine, 2007, p. 3)

Within Foucauldian approaches, subjectivity replaces identity to capture the idea of simultaneously positioning one's self in discourse and being positioned by discourse, both subject and subjected. Power is intrinsic to this process because it 'traverses and produces things, it induces pleasure, forms knowledge, produces discourse. It needs to be considered as a productive network which runs through the whole social body, much more than as a negative instance whose function is repression' (Foucault, 1980, p. 119).

This doubleness to subjectivity is important for thinking about choices because often people choosing are seen as either active makers of meanings, consciously and rationally choosing subjects, or passive consumers of meanings, subjected to the undue influence of the media and other people. However, we see choices as always and inseparably both active and passive. Neoliberalism may ostensibly rely on the rational conscious active individual, but mass consumption was constructed by playing 'on the unconscious desires of the people and associat[ing] those with the purchase and ownership of commodities' (Walkerdine, 2007, p. 7). Thus, this approach cuts across oppositions through which we usually think the construction of people: social/individual, unconscious/conscious, inside/outside, passive/active, fantasy/reality, irrational/rational, connection/separation, relational/autonomous, object/subject, body/mind, and other/self.

CONSUMING YOURSELF

We asked participants to discuss how they made subject and career choices, prompting them concerning the influence of teachers, family, and popular culture. However, the most common reasons given for their choices were unprompted ones: identity, enjoyment, and ability. So choices are constructed as a way of realising yourself. Although there is an important distinction between reasons given for choices and other ways of understanding how choices come about, we are interested here in the discourses through which choices can be legitimated. Alongside this investment in choices as self-realisation, it was common to play down the role of external influences. For example, participants spoke about discussing their choices with their parents and being supported by them, but simultaneously asserted that they were their own individual, independent acts. Even three people who want to work in popular music said that popular culture had not influenced their choices. We argue that these influences are denied because they threaten the production of the autonomous self.

The three central factors of identity, enjoyment, and ability are interwoven in different configurations in all participants' stories. We begin with examples from school students:

> *Chantz:* [Maths is] not a thing that comes completely naturally to me. If I'm going to do subjects, then I'm going to have to work hard, I should work hard at something that I'm good at or something I enjoy. Not just slog on at something I hate. [Female, White English, working class]

> *Wilbert:* I've been generally quite good at English through school. So, you know, that's always been a strong point in my education, I really enjoy it. And music itself, I'm really good at, so I don't know, and I enjoy it as well because it's something that I do. And I'm interested in the technology side of stuff. And I took media studies because it ties in with English nicely, and learn about the media and everything. And I did photography because I've never done it before but I think I will enjoy it because I'm more of a creative person. [Male, White English, middle class]

Enjoyment and ability are tied together in complex ways. Both can function as a foundation for particular choices. In Wilbert's talk, we can see a link with identity, where being a particular type of person, 'creative', is used to predict future enjoyment and as a guarantee for choices, even of subjects in which he has no experience. 'Creative' is part of a powerful discourse that maps onto and into a particular set of subjects. While there is some flexibility, our own and other research has found a strong discursive opposition between mathematics and creative subjects (Bibby, 2001; Mendick, 2006). In the next extract, enjoyment and ability are again tied to being or becoming a particular sort of person. Deebo has to be able to occupy a subject identity in order to be successful at it:

> *Deebo:* Well these [information technology, business studies, geography] are all subjects like I personally enjoy a lot. [. . .] It has to be enjoyable, yeah, coz it's like business studies as well, what I do is like I visualise myself in a situation and then I, I think like, 'oh so what would I do if I was—, knowing what I'd been taught?' So that's how I answer my questions, it's the same with geography as well. [Male, Black African, working class]

In the following extract, Pisces speaks about continuity between his subject choices and his life outside school:

> *Pisces:* I just thought it was important that the subjects I took were things that were important to me personally. [. . .] I'm an English man really, I'm into the creativity and looking at other people's work and that kind of thing.[. . .] Anyone can write English, anyone can express their own feelings in a way that anyone else can engage with. And I write quite a lot, and so it's like, it's just interesting, you're just, it's like you're improving yourself [. . .] I've always liked science [. . .] I can't really, physics doesn't really relate to me as a person, you know? It's, it's there, it's around you; it's forces and motions; but what does it mean really? What relevance does it have, to me, to my life? Whereas biology and chemistry, you can see it, you can use what you learn in your everyday life and you can interact with it. Coz it's something that can interact back with you. [Male, mixed heritage Black African and White English, middle class]

His ambiguously phrased claim to be an 'English man' ties his self to his subject. Both Deebo and Pisces talk about interacting with subjects, creating and improving their selves through this. These are just a few examples from many; all come from school students, but the pattern recurred among university students; as one social science undergraduate said, 'if you enjoy something you want to do it'.

Enjoyment and ability, linked by identity, take on a foundational role, carrying the promise of good results, good jobs, future success, and future happiness. The way that people now choose for enjoyment, ability, and identity relates to shifts in how we think about the self associated with neoliberalism. Anne Cronin (2000) traces the development of modern conceptions of the self showing how it 'gradually comes to be seen as a repository of unique and authentic potential which must be realised through processes of discovery, expression, and thus self-actualisation' (p. 30). In contemporary Western societies, this manifests as 'a process of active 'choice' framed in terms of consumerist engagement with the idea of self as project', which is then 'construed as a personal duty and responsibility'. Beverley Skeggs (2004) has shown that this version of the self is classed and gendered:

> 'The individual' is defined through *his* capacity to own property in his person. He is seen to have the capacity to stand outside of himself, to separate 'himself' from 'his body' and then to have a proprietal relation to himself as bodily property. [. . .] These different relationships to property—in objects in the person—have resonance today with those who can invest in themselves and those who cannot. (pp. 9–10, italics original)

The rendering of choice as personal responsibility means that making a 'wrong' choice is constructed as an individual failing. This ignores the ways that the parameters for choice and even the possibilities for occupying the position of the autonomous consuming self are constrained by conditions of poverty, social class, gender, and race. It blames people for their own oppression.

CONSUMING MATHEMATICS

We shift now to ask: How can choosing mathematics become part of a project of 'compulsory individuality' (Skeggs, 2004)? This question is interesting given the way that we found that mathematics is constructed as having little room for self, being seen as rigid with clear answers (see also Burton, 1995). Earlier work on choosing mathematics at 16+ found that reasons most commonly featured future job requirements or a desire for intellectual kudos and enjoyment featured in relatively few (Mendick, 2006). To explore the ways that people consume mathematics, we turn to our 11 participants who had chosen mathematics:

Alice	Female	Turkish Cypriot	Post-1992	Middle class
Boris	Male	White English	Post-1992	Intermediate class
Bridget	Female	White English	Russell Group	Middle class
Dave	Male	White English	Post-1992	Working class
Dave RG	Male	White Welsh	Russell Group	Working class
Elizabeth	Female	White English	Russell Group	Middle class
Joanna	Female	White Welsh	Russell Group	Middle class
Mr 37	Male	White English	Post-1992	Middle class
Ricardo	Male	Black African	Post-1992	Working class
Robert	Male	White English	Russell Group	Middle class
Sophie	Female	White English	Russell Group	Working class

For this group, too, the factors of enjoyment, ability, and identity come out strongly in their talk about choice, featuring in various configurations. For Joanna, her loss of enjoyment in mathematics coincided with struggles with the subject in the first 2 years of her degree, so now she mentions enjoyment only in relation to other areas, such as environmental issues. Bridget, Dave, Dave RG, Elizabeth, Robert, and Sophie all spoke about enjoying mathematics and having ability at it. For example:

> *Dave:* But with maths, I was learning how to do things and everything I was doing I was doing well at. So I just found it was the thing I was skilled at and it was the thing that I quite enjoyed being skilled at.

His continuing with mathematics is constructed not as an active choice, but as finding and expressing his self: something that he always already was.

Ricardo discusses his desire to pursue financial mathematics in the future by imagining his self doing this and the pleasure, excitement, and fulfilment this provides:

> *Ricardo:* I can imagine myself one day predicting the movement of the stock, and then of course I know I will make a fortune out of it but the, the glory that will come with it would make me feel like I fulfil something. Fulfilment is the most important part. So even though I earn money as well but I just want to enjoy it. I don't like doing something I don't enjoy.

Glory, power, and mastery are also part of the attraction for Ricardo (Walkerdine, 1988). Interestingly for Alice, her pleasure in mathematics is mediated through teaching. She does not seem to be able to occupy the role of mathematician, seeing women more often in the role of servicing and reproducing the subject than of creating and mastering it, 'not producing any more ideas or whatever, but, you know, maybe helping some other areas'. Mr 37 says that initially his choice of mathematics was due to 'laziness', having a 'basic aptitude' for the subject and encouragement, but he feels it would be better for him to have had a stronger motivation based on what he wanted to do, one which he now has as a mature student. Thus, the constellation of enjoyment, ability, and identity allows choosing mathematics to become a route to self-actualisation for many of these participants, but there are important differences between the way that story plays out in relation to mathematics compared to other subjects.

For mathematics, ability and enjoyment are more closely linked than for other subjects; for Joanna and Dave, discussed earlier, they feel interchangeable. Joanna lost her enjoyment in mathematics at the same moment that she was no longer able to do it. A further difference is the way that, for mathematics, ability is constructed as a more important foundation for choices than pleasure:

> *Boris:* If you're strong at maths, then you'll continue with it. But if you're weak at it, even if popular culture encouraged it, you still won't go into it because you don't feel you're good enough for it.

For mathematics, ability produces enjoyment in a relationship that is the reverse of that usually constructed for other subjects (see Chantz, quoted earlier, for whom enjoyment enables hard work which leads to success). This different relationship between ability and enjoyment is what enables mathematics to be inscribed as a truth about the self that can be realised through choosing mathematics. However, it means that this choice is not an active work of self-creation; they are more chosen than choosing.

The identity claims around ability and enjoyment often make use of the word 'always'. This applies to many subjects, but happens more often with mathematics in our data:

> *Sophie:* I've **always** liked it and that's why I've **always**. [You like the certainty of it?] Yeah, I think that's—that's probably why it's **always** been my best subject because I've enjoyed that part of it. The thing is like, science and things at school, I **always** enjoyed a lot more than anything like English. Purely because I was better at them so I put more effort in, really.

> *Dave RG:* I've **always** seemed to enjoy it, like at primary school, we used to get sets of the maths books and there used to be a great feeling of finishing first, getting everything right and just, it was **always** about being picked out, good at that area, and it would just make you feel really good and it's just—maths just keeps making you feel good really.

The insistent use of 'always' is striking. This word, as Judith Butler (1997) argues in the context of gender and sexuality, functions to inscribe the feelings attached to it as essential characteristics of the self and to disappear the process of coming into discourse, being assembled as mathematically able. But if choosing mathematics is to be a route to self-realisation then, in the context of the large-scale absence of other ways of interacting with mathematics, the subject needs to be experienced as an essential part of the self. It has to make you feel different and special when realised or, as Dave RG puts it, to 'just [keep] making you feel good'. There were other evocative examples of this offered in response to our request for memories that were important to participants' relationship with mathematics. For example:

> *Joanna:* When I was younger, starting from primary school, I **always** enjoyed maths and the teacher's **always** telling me how good I

was. And then in secondary school, I won—there was like a maths competition in Wales and I got the highest mark, and I just remember the day I found out. My maths teacher was so excited over me. And I just felt that was the first thing I'd achieved, like something big I'd achieved. So I suppose that's when my interest in it sort of grew.

In Joanna's memory, as in Dave RG's, we can see how recognition by an/other is crucial to this feeling of being special.

However, this notion of specialness is not innocent. Discursively, it relies on the idea that not everyone can do mathematics and that those who can are born with this 'ability'. So while literacy is seen as an essential part of being fully human, 'in contrast to this framing, arithmetic is not naturalized as genetically human, but as *genetically determined within humans*' (Damarin, 2000, p. 76, italics original).

In our data, those both outside and inside mathematics do work to maintain the boundaries between mathematical and nonmathematical people. This is particularly so for the undergraduates who have made clear choices either towards or away from mathematics. In the remainder of this section, we look at three examples of these boundary-maintenance practices. The first example, 'othering', came out more strongly in the focus groups than in the individual interviews, suggesting that it is supported by collective subject cultures. It is exemplified in these extracts from social sciences undergraduates:

> *Ellie:* I think all the social studies students probably have a natural aptitude for this and I think that there is kind of like the intelligence where you would make up an argument and you rationalise like different arguments. And then there is kind of intelligence where you would solve a problem and there is a distinct answer. And I think people just find one or the other type is easier.
>
> *Sky:* I know it's a bit over simplistic, but like if people do have areas of their brain that they are more sort of successful with. Just like looking between like sociology and maths like we are better at sort of constructing an argument and that sort of thing, better at theorising. Or maybe just because we have to be but maybe you could argue that we are more used to engaging our emotions and things like that. And so I suppose you could say with some people if their main focus in their brain is on the like mathematical technical side then that might be at the cost of the like emotional.

Othering partly works through marking the privileged mathematical category as undesirable. For example, Sky suggests that those doing mathematics are less able to engage with emotions. Similarly, 'geek' discourses, which came up often in our data, associate being mathematical with social incompetence. However, it is important to focus on the ways that these devices maintain the boundaries between the mathematical and the rest, the higher status of mathematics, and so the operation of power (Moreau, Mendick, & Epstein, 2009).

A marked example of boundary maintenance by the mathematics undergraduates is their responses to our request that they imagine a world where mathematicians appear on television regularly. We expected they would welcome this, and indeed some did, feeling it would be nice if mathematics were better appreciated. However, some did not or expressed reservations. In these cases, we can read a desire to maintain their difference from others. Sophie likes our current world as she does not mind being seen as a geek, is amused by it, and likes shattering the stereotypes. Elizabeth says, 'Maybe I quite like the fact people say with maths, they're different'. Dave RG says, 'It feels good [...] you're one of the few people who can do it'. Bridget worries that there will be more competition for university places so she will lose her place at an elite institution. Boris and Alice talk generally about how people would dislike it, perhaps wanting to maintain their difference from others.

A final example is the mathematics undergraduates' views on Carol Vorderman, who does mental arithmetic on *Countdown*, a popular UK television game show. All but one of them reacted against her. Four of them target her mathematical abilities. Sophie, Joanna, Bridget, and Mr 37 do not like the fact that she represents mathematics in the popular media, arguing that just because she is good at 'mental arithmetic' does not mean that she is good at maths. Bridget points out that she 'got a third [class degree]'. Mr 37 says:

> I don't know how much of a good mathematician Carol Vorderman is, I just know she's a good, she's good at arithmetic. [...] I think a lot of mathematicians I know state plainly that they actually can't do arithmetic. [laughter] [...] it's not necessarily about sort of deep mathematical ideas.

These kinds of oppositions between activities of calculation and mathematics, between calculating and reasoning, between surface and depth, and their projection onto people are important for maintaining the elite position of mathematician (Mendick, 2006; Walkerdine, 1997).

CONCLUSIONS

In conclusion, we return to our earlier questions: What makes it possible for us to think of someone as making a choice (not) to do mathematics? What effects does this have? We have shown how to see someone as making a choice at all relies on our ability to construct them as a particular kind of self, for 'life has become a skilled performance' (Rose, 1999, p. 242). We are critiquing the idea of choice in order to recognise how neoliberalism works to produce subjects who have to work with the 'serious burdens of liberty' (Rose, 1999, p. 12) and how some people are excluded in this process.

To make yourself through a subject choice, you must be able to construct your identity in relation to the chosen subject. We argued that discourses about feeling that you can interact with the subject, that it relates to your daily life, and that it represents something you want to become are ways of doing this. However, since, for most people, mathematics is viewed as rigid and external to the self, these discourses are not available to them when explaining their choice to do mathematics, though the absence of them is a very available route for explaining why someone did not choose it. We argued that what comes to substitute in many cases is the way that identification with mathematics is seen to confer difference and specialness onto the chooser. However, in using this discourse of specialness, doing mathematics becomes for them an expression of their inner essence and so is less choice than destiny. Thus, pursuing mathematics does not fit easily into a neoliberal framework.

We reflect now on the effects of this analysis. First, there are implications for an education policy based on choice which is likely to mean that only those successful at mathematics (and perhaps lacking confidence in arts subjects) are likely to choose it. Those who do choose it are vulnerable to moving away from it as soon as their ability to do mathematics is called into question. Second, since, as we argued, specialness is based on discourses that construct boundaries around the mathematical, this excludes many from a powerful area of the curriculum and from a range of jobs, resources, and positions to which mathematics controls access. Specifically, since ability has been shown to be classed, gendered, and raced (Gillborn & Youdell, 2000; Jones & Myhill, 2004), this reproduces existing inequalities.

The idea of specialness finds support within mathematics teaching practices and beyond, where boundary maintenance work is constantly enacted. In mathematics teaching, the use of competition and grouping by 'ability' support discourses of specialness. Beyond that, in a recent example, the U.S. television series, *Heroes* (Kring, 2006), tells interlocking stories of a group of people who have a range of special abilities. Their abilities are linked directly to evolutionary progress, and these special people save the world.

However, not all our mathematics undergraduate participants talked about themselves as special. Specialness did not feature strongly in any of the mature students' stories. This group of students are more likely to have mathematical identities that they have acquired, whereas the younger students are more likely to have ones that have always been there. For example, in this extract from a focus group, Thierry constructs mathematics as empathic:

82 *Heather Mendick, Marie-Pierre Moreau, and Debbie Epstein*

> *Thierry:* When you understand mathematics, the why and the how, then you can understand life. [. . .] Your idea is right but I have to understand I have to know why you are telling that, why you are doing this. [. . .] If you are explaining very well I won't be angry. That is why people have told me that I am so quiet, that is right, because I am always thinking. [Male, Black African, middle class]

We do not have space to develop all these stories here, but mention them as a reminder that no discourse is totalising. Their stories can offer clues to how to develop alternatives to specialness and thus more inclusive ways of being mathematical in these 'new times'. In particular, we would welcome a pedagogy based not on differences between those who are mathematical and those who are not, but on differences within the range of ways of being and doing mathematical.

ACKNOWLEDGMENTS

Big thanks to everyone who supported the work of this project, especially the participants, members of the advisory group, and the funders: the Economic and Social Research Council (RES-000-23-1454) and the UK Resource Centre for Women in Science Engineering and Technology. Also thanks to Anna Llewellyn, Cathy Smith, Jocey Quinn, Sumi Hollingworth, Valerie Walkerdine, and Yvette Solomon, who provided useful feedback on earlier drafts of this chapter.

NOTES

1. You can view examples of her work by searching for RachelWhiteread at http://www.tate.org.uk

Part III
Curriculum

As Jenny Shaw (Chapter 8) points out in the first chapter of this part, the word 'curriculum' originates in the racing tracks of ancient Greece, and it denotes a course, something which is planned and 'delivered' with an outcome in mind. It extends beyond a collection of materials: it includes the choice of those materials and the way in which they are put into practice; it is experienced by learners in terms of what and how they study and what they think they are supposed to do; and there is, of course, the hidden curriculum, the learning that is never taught. In addressing the question, 'How does the nature of mathematics curricula impact on learner identities and relationships to mathematics?', the chapters in this part are united in their focus on the role of words and symbols in the way that the curriculum is experienced and enacted. Thus, Jenny Shaw focuses on the raw emotions triggered by mathematical words and symbols, while Candia Morgan (Chapter 9) draws our attention to the discursive positioning of the learner as adept or not, powerful or not, and participatory or not in the words of government policy and in textbooks. Finally, Birgit Pepin's chapter (Chapter 10) on the use of textbooks as tools in the figured world of the classroom shows how important it is to consider curriculum in practice.

Jenny's chapter begins with the observation that, in Britain, as in many other countries, many people say that they do not like mathematics, and they say so in increasing numbers and at increasingly younger ages. In an attempt to explore why this might be, her chapter uses psychoanalytic theory to unpack the nature of like and dislike in mathematics, and the ways in which its language may engender and support the strong feelings which are so often associated with it. As she points out, mathematical words are loaded with meaning, connecting with 'power, glory, negativity and destruction'; while they may have unambiguous mathematical meanings for some, words such as 'cancelling', 'mean', 'power', 'negative', and 'taking away' have other, darker, connotations. Jenny also draws our attention to gender as a (not so) hidden meaning of mathematics—frequent references in schools and elsewhere to its 'hardness' compound its association with valued stereotypically male characteristics of power, authority, and so on. Right or wrong, and all or nothing, the hierarchical nature of mathematics and the assumption of mathematical competence as a pure measure of intelligence creates danger: as Jenny says, this image means that the possibility of future failure accompanies every triumph. Associated with these hidden meanings, she argues, is a disciplinarian teaching style epitomised in a testing system which aims to 'stretch' learners and place them constantly at risk of being *wrong*, a constant 'all-or-none' reminder of the possibility of loss. Turning to the symbolism of numbers, Jenny uses the famous case of 'Sean's numbers' to illustrate her idea that numbers can engender feeling—perhaps in family histories or early experience. She argues that these experiences can leave an association which is personal and affective, predating conscious cognition, so that 'twoness' can take on particular significance for children experiencing their parents' relationship breakdown, for example. Drawing on object relations theory, she suggests that the origin of our strong feelings about mathematics lie in our early experiences: learning in psychoanalytic terms can be seen as a process which has its prototype in the mother/infant relationship and the infant's projection of bad feelings on to the mother, who modifies these to a more 'digestible' form. The mathematics curriculum as deeply 'indigestible' can then be seen in terms of a 'wilfully misunderstanding' object which is never modified by the teacher into a more acceptable form.

How mathematics is in fact constructed by discursive practices is the subject of Candia's chapter, which draws our attention to the fact that, in the mathematics curriculum, and contrary to what may be commonly assumed, 'choices about mathematical content and the way it is to be presented to students are not based solely on logical considerations'. Her use of Critical Discourse Analysis differs from the other discourse chapters in this book, enabling her to present a close-grained analysis of the ways in which the grammatical construction of official texts position mathematics students in particular ways which delimit *what* they may say and *who* may say it. Although this is not a deterministic process—there are, of course, alternative discourses which individual teachers and learners may draw on in constructing their mathematical identities—Candia argues that the official discourse is privileged over others, fed as it is by regulatory practices such as examinations and league tables. She presents two very different examples of this discursive positioning: the first concerns the shaping of student roles through education department guidance to teachers, and the second concerns the shaping of mathematics itself and the student's relationship with it via textbooks. In the first of these, we can see how the language used to depict 'good teaching' positions students as the recipients of questions (never initiators) or as divided into differing abilities, with associated differing relationships with the teacher and each other. In the second, Candia's analysis of two textbooks—one aimed at 'intermediate-level' GCSE students, one at 'higher level' students—contrasts the extent to which the student is constructed as having agency in mathematics. As we might predict, the intermediate text positions the student as unable to question or participate in the mathematics or to challenge the authority of the writer; the higher level text, however, indicates that there may be uncertainty in mathematics and positions the student as a potential actor in its practice. Challenging the legitimacy of these differences and the discourse of ability which they rest on is possible, but only when they are laid bare by this kind of analysis of how words function to convey a hidden message about expectations and relationships.

Finally, Birgit's chapter provides a further layer of analysis which shows how the figured world of the classroom and the roles that teachers take up within it mediate the messages of textbooks, producing particular experiences of the curriculum and, in turn, different learner identities. What makes her contribution particularly illuminating is the contrast she is able to draw among different countries, both in terms of their use of textbooks and of the teacher's role with respect to them. She begins by analysing the role of textbooks in mathematics classrooms in Germany, France, and England, drawing our attention to the correspondence between textbook types and setting in England and a parallel process of different books for different types of school (technical, 'secondary modern', 'grammar') in Germany, and the sharply contrasting French practice of mixed-ability teaching and a philosophy that all students are 'entitled' to the same curriculum. She reports on further differences in teachers' use of textbooks: in England, teachers tended to rely on them heavily to support individual student practice after teacher presentation—pressurised by time, they sought 'off-the-shelf' teaching schemes and supported the practice of different textbook types for different sets. In both Germany and France, however, teachers used books far more selectively in their lessons, preferring to emphasise challenging whole-class discussion and conceptual understanding supported by their own teaching; they also upheld the benefits of access to a range of difficulty in problems for all students. As Birgit suggests, these contrasts enable us to see how some things that we may take for granted need not be as they are—in terms of our own figured worlds, they create opportunities and frameworks for refiguring classroom relationships with mathematics through different enactments of the curriculum on the page.

8 Appetite and Anxiety
The Mathematics Curriculum and Its Hidden Meanings

Jenny Shaw

INTRODUCTION

Like most enthusiasts, those who love mathematics are keen to share their joy and disappointed when few leap at the chance to do some quadratic equations or dynamic geometry. A problem mostly affecting teachers, it is also a worry for governments, including that of Britain, keen to keep ahead or even just abreast in science and technology. Views on the causes of this problem vary, from cranial 'hard wiring' to the software of 'culture', including national culture, and several points between. Although the flight from maths is much lamented, there is also a widespread feeling that those who can and those who cannot do mathematics are almost different species. Mathematician John McLeish (1991) calls this the 'binary assumption', while Alan Cane (2007), in a review of some popular books about mathematics, doubts the feasibility of such an exercise and concludes, 'Those willing and able to be excited by mathematical ideas will live with the equations. The rest are a lost cause' (p. 31). From each side there is incomprehension: those who can do mathematics cannot understand how anyone else can fail to do likewise, while the permanently bewildered are unable to imagine what it might take to be good at mathematics. Still, the 'binary assumption' cannot account for the **declining** interest in mathematics, or the fact that in Britain children, at ever younger ages, are claiming that they do not **like** the subject.

Liking, or not liking, is a feeling, and feelings, despite much ignorance of their causes, affect the choices we make, including whether to stay with or reject learning about mathematics. What exactly lies behind our feelings is often a mystery, but it is precisely because the psychoanalytic approach, generally speaking, seeks to understand the mystery of our feelings, how they work, and how they affect and interfere with actions or desires, including a wish to learn mathematics, that it might shed light on the defection from mathematics[1]. Psychoanalysis is a theory of meaning and, more particularly, of hidden meanings based on early life experience and various feelings which stem from that time. However, the psychoanalytic approach is not one commonly encountered in mathematics education, partly, perhaps, because it is associated with therapy and malfunctioning and, partly, because it invokes the concept of unconscious processes which do not lend themselves very easily to empirical observation and testing. But if you listen to parents and teachers talking about children and their choice of subjects, you will often hear comments such as, 'I don't know what **stops** Freddie from doing or continuing with—whatever subject'. Rarely is intelligence questioned, but there is a suggestion of something obscure and unseen at work, some obstacle, hidden meaning, inhibition, or a phobia—in short, the stock in trade of psychoanalysis. We have to be careful not to treat any decision not to study mathematics as somehow inexplicable or pathological. There are loads of good reasons for preferring to spend one's time doing something else. But where there is a general pattern, as there is with the flight from mathematics, there is a possibility of a more general force at work. The first possibility I want to suggest is that the language of mathematics, which might be off-putting, carries a horde of hidden and some not-so-hidden meanings relating to power, glory, negativity, and destruction.

MEAN, LEAN, AND HARD, OR POWER, PERSECUTION, AND PUNISHMENT: THE 'SHADOW SIDE' OF MATHEMATICS

The meanings of some words used daily by mathematics teachers, such as cancelling, killing, subtracting, and taking away, may seem clear to them, but for many more people, less familiar with mathematics, other meanings jostle around and possibly interfere with the process of learning. Many words have several meanings, including 'mean' which may refer, in a mathematical context, to something middling, but even then, and in a mathematics class, most teachers would clarify and distinguish between mean and modes, medians and averages. But 'mean' can also mean something lowly, and it can be used to clarify, as in 'What I mean is', or it can refer to some spiteful act or intent to harm. In the classroom, this last meaning is less likely to be discussed than any of the others, but the negativity associated with the word is still there, haunting it and making us feel uncomfortable. Whenever we clarify a word, in whatever context, we show our awareness of possible misunderstanding, but rarely do we risk, and perhaps certainly not in classrooms, going into the darker, or unconscious, meanings of words and behaviour. However, psychotherapists do, and the example given by Gianna Williams (1997) of how an adolescent patient reacted to the word may make this clear. A young man who was always very observant about time, who used it to attack his therapist, and regularly complained about the firmness of session times, one day triumphantly pointed out to her the **mean**-ness of Greenwich 'mean time'.

Of course, more 'creative' words are used in mathematics, too, adding and multiplying, for example, but the negative terms, including 'negative numbers', outnumber the benign ones and possibly have a deterrent effect. Even the word 'curriculum' may spook mathematics. Although today it means a course of study or a collection of subjects, it originally meant a race chariot, and a trace of this meaning survives in the sense of the curriculum as a hierarchy. Within that hierarchy, the position of mathematics at the pinnacle, and as the ultimately all-encompassing discipline, is unassailable. Piling it on, in ways which may become, or feel, personally dangerous, is the term 'poly**math**', for someone who is good at everything, and the practice of treating mathematical ability as a near 'pure' measure of intelligence. For with glory comes the possibility of the fall, and, as in all winner-take-all situations, most of those who enter the field of mathematics (which feels like a race) will, sooner or later, feel themselves to be failures. This 'outcome' is made abundantly clear in chapter 3 of this collection, where the writers report on how they had shared their intellectual biographies and discovered that they **all** felt like failures as mathematicians. Now if this is true, something is seriously amiss, because if objectively high-achieving teachers do not know, personally, what it means to be good at mathematics, they cannot offer models of success to their students. Not everyone can be a Richard Feynman, but if teachers feel that anything less counts as failure, this will be transmitted to their students.

The grip which mathematics has on intellectual self-esteem affects not just mathematicians, but many outside the field who admire mathematics, want to be part of it, and then beat themselves up about not being good enough. This process is neatly illustrated by, of all people, the psychoanalyst Wilfrid Bion (1970), who, in a preface to the index of his book, *Attention and Interpretation*, confesses to his frustration/despair in not succeeding in using mathematics as an idiom for expressing his ideas about the nature of thinking. Bion had intended the book to be an explanation of psychoanalysis for the lay person, and in it he proposes that the development of conceptions is analogous to the development of mathematical elements. But he is not confident, and in this confessional preface he writes of how both book and index were 'the outcome of an attempt at precision (and of how) the failure of the attempt will be clear'. Then he adds, 'what may not be clear is the following dilemma; 'precision' is too often a distortion of the reality, 'imprecision' too often indistinguishable from confusion'. Why precision was so desirable, and why reality, which is never precise, was so disappointing is another story. Perhaps Bion was alluding to debates and tensions in psychoanalysis as others in the field, for example, Donald Winnicott (1971), saw confusion/imprecision in a more positive light and lauded it as critical for the growth of autonomy and as fostering a sense of creativity. Both Winnicott and Bion are remembered for theorising the mother/infant relationship, but while Winnicott was openly sympathetic to feminine virtues, in Bion's yearning for the legitimacy of mathematics, there seems also to be a yearning for the security of identification with masculinity. Gender is one of the not-so-hidden meanings of mathematics, and the reputation of mathematics as a 'hard' subject comes partly from the fact that it shares some of the same characteristics for which men are valued—that is, of being potent, precise, authoritative, determined, demanding, and willing to take the grand view.

Mathematics is one of the most masculinised of subjects, at least in Britain, both in terms of the gender of those studying it (or related subjects such as information technology, physics, and engineering) and the stereotype of it as an entity. The two, of course, are mutually reinforcing: subjects seen as masculine attract boys and men and deter girls and women. There are plenty of reasons for this polarisation around gender, the gender and personality of the teacher; who one's friends are, and what subjects they do; having a particular career in mind; or simply whether the moment at which decisive choices have to be made coincides with a developmental change where there is intense internal and external pressure to establish personal and sexual identity.[2] But these must be skimmed over if a focus is to be kept on how the inner world and unconscious fears and feelings affect the doing, or not, of mathematics. However, one of the corollaries of mathematics being associated with men and masculinity is that it is also associated with power, punishment, and persecution. There is an unforgiving 'Old Testament' finality about mathematics, an un-negotiable sense of being right or wrong, of being a success or a failure, of life or death, and of there being nothing in between. The mathematics curriculum shares some of these features and is often experienced as demanding and very prescriptive, especially over the order of what is learned. Defined as the abstract science of space and quantity, the power of mathematics to deal with enormity may be a comfort for some, but it can also represent a terrifying form of power and omniscience which adds to the sense of the subject as *unfeeling* and impersonal.

Feelings and associations make a difference. Although it may pain mathematicians to see their subject viewed in this way, mathematics does seem closer than most other subjects to an older, more disciplinarian model of teaching and learning which featured rote learning and teachers, mostly men, who were very stern. In reality, this is not the case, and mathematics education is as progressive as any other subject, and even times tables are no longer drilled into kids. But traces of the older, harsher regime survive; for example, it is still standard practice when constructing mathematics tests to remove items which most candidates can do quite easily on the grounds that these do not 'stretch' the candidates. This may be so, but tests also prioritise ranking over learning, and removing the 'easier' items also removes a comfort zone which can reduce fear of the test and may stem the drop-out rate. Any subject which uses tests extensively risks being felt as persecutory, vindictive, unfair, and uncaring, and even more so when that testing is done by strangers or on line, as it increasingly is in mathematics. Even more, as doubts are raised about the accuracy of online marking, both fear of the test and disrespect for the subject grow. Defenders of the test may say that tests improve performance, but the more they are used, the more those taking tests face the risk of failure and 'being wrong'. The same problem, I suspect, afflicts modern languages, which is another subject afflicted by 'flight' and which also makes much use of tests. Mathematics may be both felt and taught in more creative, elastic ways at higher levels, but the sad truth is that few and fewer students reach those levels.

The sudden death which the test represents returns us, or does for some of us, to the early life existential fears and anxieties, sometimes called 'a nameless dread' which at root is based on whether Mum is there, in which case all is well, or she is not, when everything falls apart. The terrifying 'all-or-nothing' experience is repeated many times in the course of a life and, perhaps, on each and every occasion is accompanied by a sharp descent from comprehension to incomprehension. If so, one consequence is to show how misleading the notion of the 'learning curve' is because much learning, particularly of knowledge, is not or is certainly not experienced as gradual, linear, and incremental, but as jerky, episodic, and beset with loss as much as gain (see Jordan, 1968). A theorem can be puzzled over many times and not understood a little better each time, but then something may 'click', the bits fall into place, and sense is made. Why this happens is, again, another story; a blockage might have dissolved, one small shift may have led to a major reordering in the mind, some feeling state might have changed, or some aspect of the external world. In the next section, I give a couple of examples where the learning or doing of mathematics appears to stumble: in the first case because, although the child is unaware of it, her attempts at manipulating numbers represents an attempt at manipulating her parents and their relationship. In this case, the flesh-and-blood parents are symbolised by an abstract number. In the second example, a boy senses, in a fifth dimension way, some properties of numbers which are invisible to his peers and teachers. The workings of the mind are mysterious, and any understanding of it is only ever very partial. But if unconscious knowledge and unconscious processes can affect how we act and what we allow ourselves, consciously, to know and do, they will affect the learning and doing of mathematics, too.

PEOPLE AS NUMBERS AND NUMBERS AS PEOPLE

In a chapter discussing what makes or stops children from learning, child psychotherapist Hamish Canham (2006) suggests, first, that there is a tension between the desire to know and the desire not to know, and that there is a particular pressure not to know about certain aspects of their parents' lives. Second, Canham argues that all new knowledge arouses hostility, threatens security, challenges claims to omniscience, reveals existing ignorance, and creates a sense of helplessness. He then adds, almost as a footnote, that the two subjects most at risk from the tension of the competing desires—to know and not know—are mathematics and biology. He does not elaborate further, but to illustrate his general thesis, he describes a young patient, a girl age 6, whose parents had separated, but were thinking of getting back together. In one session, the girl decided to do some drawings of faces and connecting lines, but not in a way which connected the mother figure to the father figure. In the same session, she decided to do some maths and set herself the sum of 1 + 46. But she struggled and kept coming up with the answer 46, which Canham interpreted as a sign of her having considerable difficulty in seeing her parents as a couple again (she did not want them to get back together) and, as a result, of being unable to see more than half the sum. The conflicted feelings about her parents' relationship had inhibited her intellectual capacity and, for Canham, perfectly illustrated a point made by Roger Money-Kyrle about the profound importance of the parental relationship on mental life. The key point is that number can, and in this instance did, symbolise both people and a relationship. The defining feature of the object relations strand of psychoanalytic thinking is not only to emphasize the importance of early life, and of the experience of relationships as formative, but also to see how an understanding of the world acquired at this time, particularly about relationships, is far reaching. The implication in this example is that the experience of the real parents and their relationship, and the girl's feelings about it, inhibited her capacity to manipulate their more abstract representations as number.

My second example is a case known as 'Sean's numbers', or sometimes 'Shea numbers', and turns on Sean sensing something about the number 10 and a range of properties and possibilities which he thought were embedded within it. Sean was a boy whose teacher happened to be part of a project producing teacher training materials by video recording as well as notes and diaries (see D. L. Ball & Lampert, 1999). In one third-grade mathematics lesson, when the class was doing odd and even numbers, Sean told the class that 6 could be an odd as well as an even number, and he drew on the board six circles, putting lines between the second and third circles and between the fourth and fifth, producing three pairs of twos. The teacher reports that while knowing that the boy was 'wrong', he was also 'right', and that he may have happened upon something quite profound about the number 6. She asks the class what they think, and they decide that Sean is wrong and try to explain why. But Sean persists; another classmate pitches in, grasps what he is getting at, and runs with the idea, but still disagrees. Sean then takes her example of the number 10 as confirmation of his proposition and as another number which can be both odd and even. The class gets excited, but is also worried that if all numbers could be both odd and even, the rug would be pulled out from under the discussion they thought they were having. A video clip of this episode has been much watched and discussed by different audiences: experienced teachers, preservice teachers, mathematicians, and educational researchers, and each group filtered what they saw through their own experience/professional knowledge. New teachers wondered if the teacher always taught like that, the research mathematicians were, at first, certain about what was right and wrong, while the teacher herself confessed to having felt conflicted during the lesson, including about how long to allow the discussion to continue. Later, the mathematicians were surprised by the differences among them over what each thought the boy knew. One thought that Sean and the class were encountering modular arithmetic, another thought that Sean wanted to make odd numbers closed under addition, and yet another thought that Sean and the class had been confused by the teacher using imprecise definitional language. One use which the researchers had made of the videos was to show that an interdisciplinary approach was essential if the complexity of what goes on in a classroom is ever to be 'captured' and that, when knowing is seen as interpretative, knowledge claims are made more tentatively. However, in the many hours of discussion, no one tried to work out what Sean might have been thinking or what might have been behind his thinking.

Because I have not seen the clips, what I am about to suggest about what Sean might have sensed is, of course, wholly speculative. But numbers have properties, one of which is that they can evoke feelings. Many people have a favourite number, believing, say, that the number 1 is lucky or unlucky for them. Many others deliberately select or avoid particular digits when filling out a lottery ticket or choosing a PIN. The boy, Sean, seems to have sensed something about the number 6. It was a gut feeling, but gut feelings have histories, too, and this feeling came from something which Sean had experienced as part of his family. As families grow or shrink, by adding or losing members, interpersonal dynamics change, and with each change of size comes different combinations, coalitions, groups, and groups within groups which are made up of even or odd numbers. A group of two (a parent and child), three (two parents and a child or one parent and two children), or four (two parents and two children, one parent and three children, etc.) all hold within them different potentialities. We are not born understanding number, nor with the capacity for abstract thought, but our early life experiences prime us to experience number personally and affectively and long before we can grasp it cognitively or abstractly. The case of Sean's numbers suggests not just that we can think of numbers as having an inner life, but also that our inner lives may affect how we think of number. I cite Bion and his preface again, when he makes the same point, writing that 'Mathematical elements, namely straight lines, points, circles and [. . .] what later becomes known by the names of numbers, derive from realisations of twoness as in breast and infant, two eyes, two feet and so on [. . .]'. Being connected to another person is often what makes individuals feel whole, and as being part of a 'two' is what makes a one, and being just a 'one' can feel like being a 'nought', the principle on which binary mathematics is based, nought and one, must strike some strange emotional chords.

FEEDING AND LEARNING: THE DIGESTIBLE AND THE INDIGESTIBLE

Tracing every step from the first days of life to taking up or giving up a subject such as mathematics is impossible, but I hope I have established some of the ways in which mathematics may resonate with early life and vice versa. When giving the talk on which this chapter is based, I decided to show a couple of slides in order to underline how a fear of mathematics might be linked to very early life experience and the world which greets the infant which can be overwhelming and frightening. At any age, contemplating something huge, say a mountain, can involve a sense of personal insignificance, paralysis, and, perhaps, a fear of being crushed. The first slide was of a tiny infant facing a huge breast looking a bit startled with an expression which could be read as, 'What on earth am I meant to do with this?' or 'Wow'. The second was of a teenager cramming a clutch of rulers, compasses, and computer bits into his mouth, an image of grim, gritty struggle and play ferocity. Of the two pictures, the second was more ambiguous, as it could be taken to show an attempt at swallowing something large and near impossible or of expelling it. What I hoped it caught was something about hardness and the difficulty of taking in ideas, as well as food, and how problems with either can arise from taking in too much or too little. Together, the two slides were meant to illustrate how feeding was the beginning of our mental as well as our physical lives and, as Margot Waddell (1998) puts it, that 'the kind of thinking which goes on in any learning-situation is based in processes for which the mother/infant relationship offers the prototype' (p. 106).

In making this case, Waddell also draws on Bion, who thought that all thinking was difficult, and that how we cope with difficulties, of any sort, but especially intellectual, rested on a decision between modifying or evading frustration, which in turn determined whether or not thought developed. The experience of frustration is fundamental to early life and, as Bion saw it, depended as much on how far a mother could tolerate her child's pain, frustration, and confusion as on how the child could. In an ideal world, the mother/infant relationship works as a sort of circuit of mutual thinking, understanding, and grasping of meaning. More particularly, if the mother is sufficiently emotionally engaged with her child that the two of them are almost merged in mind as well as body (a state Bion called 'maternal reverie'), she could do some of the thinking/feeling **for** her child. If she can tolerate the child's distress, confusion, and frustration and she can take in her child's feelings, hold them awhile, and modify them, the child might then be able to take them back in a more bearable form because they had been shared, and move on from confusion, frustration, and so on into another, easier state of mind. Looked at more closely, the negative feelings, or state of mind affecting the infant, is projected into and held by the mother until the child is ready to take them back, to 'reintroject' them from the mother. Of course, this is not what all mothers do, or do all the time, and often the child's negative feelings are repelled rather than accepted, modified, and returned, in which case they *lose* meaning and become what Bion calls a 'nameless dread'. Still, because he has no real choice, an infant will go on trying to get the mother to take in the bad feelings and will still projectively identify with her, even if she is not receptive and understanding. Thus, as Bion saw it, the infant is first presented with and then takes in what he called a 'wilfully misunderstanding object'. Something similar, perhaps, occurs in mathematics settings where some students experience the subject as unresponsive and 'wilfully misunderstanding'. Certainly this is how many students experience academia *per se* and, in particular, the academic jargon which they are expected to use when they arrive at university. As in the feeding context, in which, when it goes wrong, the infant can feel attacked by what it has bitten off and then respond by attacking the hand (or breast) which feeds it, so the student may attack or reject the teacher, the book, or the whole discipline which is feeding them the dreadful stuff. This ridding oneself of an uncomfortable state by locating it outside the self is called 'projection', and it is a common enough tactic and visible in the blaming of something or someone else. But the key point is that the lesson which feels that it has 'gone wrong', where the teacher cannot seem to excite or get the student to share her enthusiasm for mathematics, and the student simply cannot get it, is a version of the break in the circuit of mutual understanding between the mother and child. Instead of mutual understanding and needs being easily met, there is mutual misunderstanding and unmet need, and the situation turns into one of envy and hostility, exactly as is illustrated in some of Tamara Bibby's transcripts in this volume. Nonmeaning dominates, and negative and toxic feelings are projected into the one who does not understand, usually the teacher, but sometimes the discipline. The student may envy the 'other' who understands something they do not and may think that this other is withholding that knowledge and perversely refusing to share it, so they turn against that person.

However, I would like to end on a more optimistic, but also more personal, note. A couple of years ago, I visited a mature student who had recently returned to study after years out of education. To my surprise, I was asked by him to read along with him a chapter in an engineering textbook which he needed to understand in order to be able to complete an exercise. Because I knew absolutely nothing about the topic, I could not see how I could be helpful. But apparently I was, and he was able to complete the task. It was not just that my being there mitigated the loneliness of his struggle, but, perhaps, it revived a sense of being held and thought about as an infant might by its mother. There may have been a sense of shared thinking, although not about semiconductors, even if that was what he thought I was thinking about. But knowing that he was being thought about may have made the task he had been set more bearable and enabled him to come to grips with the material and solve the problem for himself.

NOTES

1. There are many varieties of psychoanalytic thinking, and the one upon which I am drawing in this chapter is known as 'object relations' theory. This is distinguished by its stress on the importance of relationships and the experience of them as formative and as shaping the person emotionally. The term 'object' generally means a person or part of a person (in infancy, it can mean the breast), and the core assumption of the school is that the search for relationships and someone to have one with is the driving force of life. This approach stands in contrast to, and a marked move away from, Freud's idea of life and death instincts and of seeking instinctual gratification as the driving force. One consequence of this 'turn' in psychoanalytic thinking has been an increased interest in children and their states of mind and in the mother/infant relationship as the first relationship. Experience of this relationship is generally seen as formative and as affecting how later relationships are experienced and approached (e.g., with the expectation of being welcomed, appreciated, and understood or of being fended off, rejected, and misunderstood). The early experience of being understood or misunderstood can be particularly profound and affects not only relationships with people, but also with institutions, schools, and workplaces, for example, and with bodies of knowledge such as mathematics.
2. For further consideration of subject choice as affected by the internal world, see Jenny Shaw (1995).

9 Questioning the Mathematics Curriculum
A Discursive Approach

Candia Morgan

As we may see when we consider the history of acrimonious disputes about the content of the mathematics curriculum, school mathematics is not neutral or value free. Whether about Euclidean geometry and 'New Maths' (Moon, 1986), long division (Hoyles, Newman, & Noss, 2001) or the comprehensive 'reform' versus 'traditional' confrontation of the U.S. 'Math Wars' (Schoenfeld, 2004), the forms that these arguments take demonstrate that choices about mathematical content and the ways it is to be presented to students are not based solely on logical considerations. Rather they involve, explicitly or implicitly, cultural values, concepts of the disciplinary nature of mathematics itself and of its role within society, theories of learning, of teaching, and of the nature of students.

These values, concepts, and theories are realised in the texts that populate and regulate the practices of school mathematics—curriculum documents, examination syllabuses and papers, textbooks, and so on—and thus form an important part of the discourse of school mathematics in which teachers and students participate. This 'official' discourse shapes the ways in which its participants may speak and act, their 'particular moral disposition, motivation and aspiration, embedded in particular performances and practices' (Bernstein, 1999a, p. 246). This is achieved not only by constraining what it is possible to say and do, but also by making available subject positions that constrain who may speak and act in particular ways.

The official discourse does not, however, determine the ways in which teachers and students may relate to mathematics and to each other. In every situation, there will be other discourses available, providing alternative sets of concepts, values, and subject positions on which teacher and student participants may draw. These discourses may be mathematical or educational, whether drawn from the teacher's experience of academic study and professional development or from popular discourses about mathematics and education disseminated in the media and through students' family or peer group experience. At times, teachers and students may also draw on alternative discourses from their out-of-school practices. Among all these discourses, however, the official is privileged in that it incorporates the regulatory processes of the education system (examinations, inspections, and other forms of evaluation of students, teachers, and schools) and thus impacts directly and powerfully on individual and collective experience within the institutional structure of the school.

It is through this process of drawing on the resources of the various discourses available within a given classroom that individuals construct their identities as teachers and students of mathematics, positioning themselves in relation to the mathematical and nonmathematical activity within the classroom and in relation to the other participants in the classroom and accounting—to themselves and to others—for the nature of their own participation. The privileged official discourse provides what may be considered 'natural' positions for teachers and students, although individuals may resist this discourse, whether because they choose to privilege an alternative discourse or to establish a more positive way of accounting for their practices. The construction of identity is simultaneously 'a matter of being 'subject' to, or taking up positions within discourse, but also an active process of discursive 'work' in relation to other speakers' (Benwell & Stokoe, 2006, p. 18). For example, the National Numeracy Strategy (Department for Education and Skills, 1999), a set of official curriculum documents regulating the teaching of mathematics in primary schools in England, provides an ideal student position that involves engaging with and seeking to understand several different informal and formal ways of performing multiplication. An individual student who responds to this expectation by insisting on using only a standard algorithm, saying 'My father says this is the right way to do it', draws on a different discourse of mathematics current in her home environment (one correct method, validated by authority rather than understanding). Although, within the official discourse, this student might be positioned as deviant or even failing, by drawing on different resources, she is able to account for her practice in a way that constructs an alternative positive identity as a good student of mathematics. Of course, this accounting may not be accepted by her teacher and may lead to conflict for both teacher and student.

In this chapter, I attempt to show how some of the curricular texts of the official discourse of mathematics teaching and learning in English schools contribute to constructing positions for students. In the next section, I outline the analytic approach, and then I provide analyses of extracts from a guide for teachers and from textbooks and discuss the ways in which they provide resources for students' identity work.

ANALYTIC METHOD

In order to examine the concepts, values, and positions available to teachers and students in official school mathematics curriculum discourse, I draw on principles and tools of Critical Discourse Analysis (CDA) (Chouliaraki & Fairclough, 1999; Fairclough, 2003). As a critical approach, this seeks to understand and explain how discourse works within a social practice and to identify ways in which it might be different, considering its effects on relationships and power struggles. It is thus well suited to address the problem of how students may be positioned and may construct their identities within the practice of mathematics classrooms. Of course, the discursive moments of this practice are made up of many kinds of texts, official and unofficial, written, oral, and multimodal. I have chosen to look at written texts, examples taken from curriculum documents and textbooks, because of their privileged status, in order to identify 'natural' student positions made available by the official curriculum. A study of how students actually make use of these and other discursive resources in constructing their identities would require analysis of a much broader variety of data.

The analysis of texts themselves requires a means of characterising and interpreting their relevant features. For CDA, this is provided by the theory and grammar of systemic functional linguistics. This theory relates the lexico-grammatical system of language to the social function it performs. In particular, any text performs three 'macrofunctions': the ideational function (constructing and representing the nature of experience in the world), the interpersonal function (enacting social relationships and identities), and the textual function (constructing the text as a meaningful component of social practice). Each of these functions is realised in text by specific components of the lexico-grammar. For the purposes of this chapter, I attend to a limited subset of these components: mainly the transitivity system (the types of processes, participants, and agency), realising the ideational function; and the modality (the presence or absence of adjuncts and qualifiers that vary the degree of probability or the expression of attitude), realising the interpersonal function.[1] These allow me to address the following questions to characterise the positions provided for students within the discourse:

What kind of activity is mathematics (as it is represented in the text)?
Who participates in mathematical activity and in what ways?
How is the student's participation related to that of other participants?
Where does authority lie in relation to mathematics?

EXAMPLE 1: CURRICULUM GUIDANCE FOR TEACHERS

As a first example, I take a short extract from the *Framework for Teaching Mathematics: Years 7, 8 and 9* (Department for Education and Skills, 2001), an official publication providing extensive guidance to teachers in England about approaches to teaching the content of the curriculum of the lower secondary school. The guidance is addressed to teachers, and its most obvious function is to shape the nature of teaching. I have discussed elsewhere the ways in which this text as a whole may impact on teachers' professional identities (Morgan, 2009). However, it also serves to construct a picture of the nature of students and their roles in the classroom.

The extract analysed here appears in the guide to the *Framework* as one of the strategies involved in good teaching under the subheading 'Questioning and Discussing'. What is meant by 'questioning and discussing' in the mathematics classroom? Who questions, who discusses, and why? This teaching strategy is glossed as:

> questioning in ways which match the direction and pace of the lesson to ensure that all pupils take part (if needed, supported by apparatus, a calculator or a communication aid, or by an adult who translates, signs or uses symbols); using open and closed questions, skillfully framed, adjusted and targeted to make sure that pupils of all abilities are involved and contribute to discussions; asking for explanations; giving time for pupils to think before inviting an answer; listening carefully to pupils' responses and responding constructively in order to take forward their learning; challenging their assumptions and making them think.... (Department for Education and Skills, 2001, p. 27)

I start by examining the transitivity system: the processes present in the text, the participants in those processes, and the attribution of agency. What kinds of things do students do? They *take part*, *are involved*, *contribute*, and *think*. These appear to be active roles, yet, looking more closely, we see that they only do these things because the teacher makes them do so:

...ensure that all pupils take part
...make sure that pupils of all abilities are involved and contribute to discussions
...making them think

Moreover, although the teacher listens to their responses, there is no suggestion that students might initiate a question themselves. Thus, while students are clearly expected to participate actively, this participation is only in response to teacher action. The only thing that students seem to do independently is to make *assumptions* that the teacher should challenge. The student constructed in this text has very limited agency—and where she does act independently, it is likely to be wrong.

A further point of interest in this text is the emphasis on participation of *all pupils*. On the one hand, this might appear to be an egalitarian expectation: No pupil is to be excluded from participation in classroom interaction. However, at the same time, the participation of at least some pupils is problematised. They will not all take part equally. Rather, the teacher must *ensure* that they do so; they may need to be *supported*; the teacher must *adjust* his or her practice for *pupils of all abilities*. Thus, the text constructs differences among students, contrasting those who may participate spontaneously with those who will not unless supported (or forced) to do so.

The ideal student is positioned as one who participates by responding to teacher initiations. The deviant student is one who fails to respond appropriately (making assumptions) or who requires a variation from the default practice (support or adjustment in the form of the teacher initiation) in order to respond. There is no space within the discourse for a student to initiate discussion; in order to do so, a student would have to draw on resources from an alternative unofficial discourse, which may or may not be recognised by the teacher and other students in the class as a legitimate form of participation.

Of course, this text is addressed to teachers and is not one with which students interact directly. However, it provides discursive resources for teachers to make use of within their classrooms, hence making them also available to students. By its impact on teachers' practices, it serves to position students, the resulting forms of pedagogy constraining the ways in which they may act. At the same time, the differentiation in teacher behaviour toward 'ideal' and 'deviant' students and the currency of the notion of ability provide the constructs *able* and *less able* (or *unable*) as resources for students to employ as they account for their practices in the classroom, thus incorporating them into their identities in the classroom context.

EXAMPLE 2: TEXTBOOKS FOR STUDENTS

As a second example, I take extracts from a pair of texts intended to be used directly by students. Practice in relation to the use of textbooks varies between and within countries, although in most classrooms, the text is likely to be mediated by the teacher, and this will affect the ways in which students interact with the text themselves. Haggarty and Pepin (2002) note that in England, while students themselves make relatively little use of textbooks, their teachers use them extensively in planning lessons (see also Birgit Pepin, Chapter 10, this volume). Textbooks thus have a strong influence, whether direct or indirect, on students' experience of mathematics. As students construct their understandings of the nature of mathematics and mathematical activity and of their own identities in relation to mathematics, they will draw, to different extents, on the textbook, the teacher's speech and actions, and on their previous experiences.

Unlike in some other countries, schools in England have considerable freedom to choose the books they will use; although almost all mathematics textbooks conform closely to the topics prescribed by the National Curriculum and by examination boards, there is variation in the ways in which these topics are presented: the forms of explanation and the types of tasks posed for students. This variation is between books produced by different authors and publishers, but also between books produced for different groups of students (see e.g., Dowling's [1998] analysis of texts intended for

102   *Candia Morgan*

high and low attainers). Each textbook not only constructs a particular image of the nature of the content of mathematics, but also of mathematical activity and the place of human beings in general, and student readers in particular, in relation to mathematics.

The two extracts that follow are taken from textbooks in the same series, written by the same authors, and intended for students in the final year of compulsory schooling (ages 15–16) preparing for GCSE examinations at Intermediate and at Higher level (national examinations taken at age 16+ at the end of compulsory education, set at different levels for students with different expected levels of attainment[2]). Both extracts present definitions of trigonometric concepts, although they presuppose different amounts of prior learning in the topic[3]. In analysing these two texts, I look, as before, at the transitivity system of each text, considering what kinds of activities are present and to what extent the student reader is constructed as an active agent in doing mathematics. I also consider the modality of statements—that is, the ways in which, through the use of modal adjuncts (may, should, etc.), adverbs, or adjectives, they express degrees of certainty and other attitudes toward events and objects.

Extract 1: Intermediate GCSE textbook (Vickers, Tipler, & van Hiele, 1996a)

> In Investigation 15:1, you found that the ratio $\frac{\text{shortest side}}{\text{longest side}}$ i.e. $\frac{\text{opposite}}{\text{hypotenuse}}$ is the same for each of these triangles.
>
> This ratio is given a special name. It is called the sine of 40° or sine 40°.
>
> The ratio $\frac{\text{adjacent}}{\text{hypotenuse}}$ is called cosine 40°. The ratio $\frac{\text{opposite}}{\text{adjacent}}$ is called tangent 40°.
>
> The abbreviations sin, cos, tan are used for sine, cosine, tangent.
>
> The ratios sin A, cos A, tan A are called trigonometrical ratios, or trig. ratios.

In this extract, the student reader is initially directly present in the text and is ascribed an active role in the recent past, having *found* results as part of a previous task—although the fact that all potential student readers are assumed to have found the same results suggests that the earlier activity was one of guided discovery rather than open investigation, thus ascribing little freedom to the student to engage independently with mathematical content. The rest of the text focuses on definition as an act of naming. However, the agency in this act is obscured by use of the passive voice; it is not clear by whom the ratios are given their names. This lack of explicit human agency distances the student reader from the act of defining. Without any indication of who is doing it, the position constructed for the student does not allow her either to question the authority or judgment of whoever has chosen these names or to imagine herself participating in doing mathematics as a person who is able to make definitions. The distancing of the student from the mathematical act of defining is further reinforced by the absolute modality of the text. Each statement is presented without any qualification, and the text as a whole is presented as a sequence of unquestionable facts.

The second extract, intended for the highest achieving 16-year-olds, not only deals with more advanced mathematical content, but also constructs a different student reader.

Extract 2: Higher GCSE textbook (Vickers, Tipler, & van Hiele, 1996b)

> The ratios $\sin\theta$ and $\cos\theta$ may be defined in relation to the lengths of the sides of a right-angled triangle.
>
> $\sin\theta$ is defined as $\frac{\text{length of opposite side}}{\text{length of hypotenuse}}$.
>
> $\cos\theta$ is defined as $\frac{\text{length of adjacent side}}{\text{length of hypotenuse}}$.
>
> Since $\theta < 90$, $\sin\theta$ and $\cos\theta$ defined in this way only have meaning for angles less than 90°.
>
> We will now look at an alternative definition for $\sin\theta$ and $\cos\theta$ which has meaning for angles of any size. [. . .] This gives the following alternative definition for the ratios $\cos\theta$ and $\sin\theta$.
>
> The ratios $\cos\theta$ and $\sin\theta$ may be defined as the coordinates of a point P where OP makes an angle of θ with the positive x-axis and is of length 1. Defined in this way, the ratios $\cos\theta$ and $\sin\theta$ have meaning for angles of any size.

As in Extract 1, the definitions are presented using the passive voice, obscuring human agency in the act of defining. However, in this case, some of the statements of definition are qualified (*The ratios $\sin\theta$ and $\cos\theta$ **may be defined***) to reduce the degree of certainty. The contingent nature of definition is further emphasised by phrases such as **alternative** *definition* and *defined* **in this way**, as well as by the suggestion that a term defined in a particular way may *only have meaning* in some circumstances. This less certain modality opens up the possibility of alternative ways of doing things in mathematics—and the possibility that the student might be able to make choices.

In this text, human activity is explicitly present only in the statement that *We will now look. . . .* There are two issues to consider in relation to this statement: what kind of mathematical activity is available to the human participant and who is it who does this? Rather than engaging in material activity and discovering mathematical facts as in Extract 1, 'looking' is an intellectual activity and, indeed, leads into a reasoned argument. *We* who engage in this activity might be construed as the authors of the text, but might also be read as including the student reader as a coparticipant. Using Rotman's (1988) distinction between the two types of roles ascribed to human participants in mathematics texts, the 'higher' student is invited to be a 'thinker' while only the 'scribbler' role is present in the intermediate text.

Comparing the two examples, we see that they construct different images of the nature of mathematical definition. Whereas definitions in Extract 1 are certain and unquestionable, in Extract 2, they are contingent and may be purposefully selected, making the student reader aware of a more powerful and creative aspect of advanced mathematical activity. They also construct different possible roles for the student reader: as one who is guided to (re)discover facts or as a potentially equal participant in a mathematical community engaging in intellectual activity. The higher level text provides discursive resources that could serve to apprentice a student reader into advanced or 'esoteric' mathematical practices (Dowling, 1998), making it possible to construct a positive and powerful identity in relation to academic mathematics—an identity that is not made available to the readers of the intermediate-level text.

There is a strong assumption in the UK school system that students with different levels of 'ability' require differentiated curricula[4]. This notion is evident in the organisation of teaching groups, the structure of the examination system, and the provision of different kinds of texts for students within the same age group, but assigned to different categories. The different relationships to mathematics available to students using the two texts discussed earlier mirror the systemic assumptions about the possible trajectories of the target groups. A vicious circle is thus constructed, in which the student who is judged to be less successful at mathematics is provided with resources that distance them even more from powerful mathematical practices.

It is interesting to note, however, that even those designated as 'more able' do not necessarily have access to resources that might allow them to construct powerful identities in relation to mathematics. In Morgan (2005), I analysed two further extracts dealing with the same topic, both taken from advanced-level textbooks intended for university-bound students (ages 17–18). Although it might be expected that, being written for similar audiences, these texts would construct similar positions for their readers, this was not the case. One text, like Extract 2, constructed an important role for human mathematicians in making decisions and, using the pronoun *we*, allowed the student to consider herself to be invited to share in this intellectual activity and to be engaged in and persuaded by argument. In contrast, the other text constructed a less powerful student role, similar to that available to student readers of the intermediate GCSE text in Extract 1. Rather than being invited to share in the decision-making activity, the student was instructed to carry out material tasks; rather than being persuaded by argument, she was presented authoritatively with a procedure to follow. Although both the advanced-level textbooks seem likely to support students and teachers in preparation for their examinations, the analysis raises questions about the extent to which students experiencing the two kinds of curriculum suggested by the two texts are prepared to participate in the discourse of mathematics at university.

CONCLUSIONS

I have offered analyses of extracts from curriculum documents, showing the resources they provide to shape possible ways for students to relate to mathematics and to mathematics classroom practices. Clearly, the student positions constructed by these texts do not determine how students may be. Moreover, the practices they suggest do not determine the practices of teachers and students in actual classrooms. As Chouliaraki and Fairclough (1999) point out, 'discourse has social force and effect not inherently, but to the extent that it comes to be integrated within practices' (p. 62). I have argued, however, that the concepts, values, and positions of the official discourse, represented in the curriculum texts I have considered here, have particular force because of the roles they play in regulating school practices and, hence, the extent to which they are integrated into the actual experience of teachers and students.

The CDA approach that I have used here has allowed a description of this official curriculum discourse. Importantly, however, it also offers the possibility of considering how things might be different. By analysing how specific forms of language shape the ideational and interpersonal meanings represented in a text, we are enabled to ask what might be possible if different linguistic choices were made and how this might impact differently on classroom practices and student identity. What different opportunities might be provided to students by making human agency explicit in mathematical practices such as defining or by modifying statements to indicate where choices have been made? How might the guidance for teachers on 'Questioning and Discussing' be written differently to construct active positions for students as initiators of mathematical discussion or to avoid pathologising 'less able' students? The discourse of official curriculum documents tends to allow little space for teachers or students to insert alternative points of view or to question the often anonymous authority of the official texts (Morgan, 2009). By identifying and questioning the assumptions that underpin these texts, teachers, students, and others involved in education may be enabled to adopt more powerful positions in relation to the curriculum and to mathematics itself, grasping agency for themselves in order to challenge the official definition of the nature of mathematics classroom practices and their own roles within these practices.

NOTES

1. Halliday (1985) provides an introduction to the whole grammar, while Morgan (2006) offers a fuller discussion of its application in the context of mathematics education research.

2. At the time these texts were published, the GCSE examination was set at three levels: higher, intermediate, and foundation. Since 2008, there have only been two levels: higher and foundation. The highest grades are not available to candidates entered at the lower levels.
3. Analyses of these texts were presented in a previous article (Morgan, 2005) to illustrate a fuller discussion of the nature of definition as it appears in the mathematics curriculum.
4. Cooper (1994) offers a powerful critical analysis of this assumption.

10 The Role of Textbooks in the 'Figured Worlds' of English, French, and German Classrooms
A Comparative Perspective

Birgit Pepin

INTRODUCTION

Over the last 10 years, I have studied curricular practices in mathematics classrooms in different countries, in particular in England, France, and Germany. The goal of these studies has been to develop a deeper understanding of what is going on in mathematics classrooms at the lower secondary level, especially with respect to teachers' use of curricular materials such as textbooks. The comparative perspective has helped to reveal teachers' practices more clearly, to deepen our understanding of those practices and discover alternatives, and perhaps to stimulate discussion about choices within our own country. Sometimes it is important to realise that we have choices and what these may look like.

In this chapter, I argue that the differences in textbook mathematics tasks, and their 'mediation' by teachers and classroom environments in England, France, and Germany, create differences in activities and practices in mathematics classrooms, and that these in turn mediate different kinds of learner identities. Furthermore, schools and classrooms themselves provide an environment where particular identities are at the same time shaped and 'available' to pupils—these are 'figured worlds' (Holland et al., 1998) where meanings are conveyed and negotiated.

CLASSROOM ENVIRONMENTS, MATHEMATICAL TASKS, AND PUPIL IDENTITY

Students spend much of their time in classrooms working with written materials such as textbooks, which appear to influence, to a large extent, how they experience mathematics (Valverde, Bianchi, Wolfe, Scmidt, & Houng, 2002). Teachers mediate textbooks by choosing, devising, and structuring student work on textbook tasks. In this sense, they also mediate student learning. For example, Doyle (1988) argues that the tasks teachers assign to students influence how they come to understand the curriculum domain and serve as a context for student thinking, not only during, but also after, instruction. This premises that tasks, typically chosen from textbooks, underpin how students think about mathematics and come to understand its meaning. Indeed, Hiebert et al. (1997) argue that students

> also form their perceptions of what a subject is all about from the kinds of tasks they do. [...] Students' perceptions of the subject are built from the kind of work they do, not from the exhortations of the teacher. [...] The tasks are critical. (pp. 17–18)

Thus, it appears that the kinds of tasks that students carry out not only help to shape their perceptions of (the nature of) the subject, but also set out the basis for the kind of instruction that they experience. Mathematical tasks are, then, central to student learning and their developing perceptions of what mathematics is and what doing mathematics entails: Tasks 'produce' particular 'mathematical dispositions' or a 'mathematical point of view' (National Council of Teachers of Mathematics, 2000; Schoenfeld, 2004), as well as developing mathematical knowledge.

According to Lave and Wenger (1991), tools (and artefacts) constitute resources, and students learn by participating in social practices using tools—in our case, textbook tasks. Textbooks become artefacts in the practice: That is, their meaning should be understood in connection with the ways they are used 'in practice'. They can only be fully understood through use, and using them may mean that the user's view of what mathematics is and what it means to do it may change. This point also relates to 'conceptual tools': If students use a conceptual tool, as perhaps advised by a worked example, a teacher's exhortations, or an exercise, and if they use the tool actively, they are likely to build an increasingly rich understanding of the 'usefulness' of this tool in their mathematical world, the mathematical world of the textbook, and the tool itself. Learning how to use a conceptual tool involves much more than the set of explicit rules it may describe. The occasions and conditions for its use arise out of the contexts of tasks and activities that students are expected to do, and they are framed by the ways the members of the community (e.g., textbook/task authors) see the world of mathematics.

Different practices in mathematics classrooms are likely to influence the development of different learner identities. For example, Boaler, Wiliam, and Zevenbergen (2000) argue that in the U.S. and English classrooms they studied, there exists an 'unambiguous vision of what it means to be successful at mathematics' (p. 8). In another study, Boaler and Greeno (2000) report that in U.S. classrooms, 'mathematics was presented as a series of procedures that needed to be learnt', and that students developed identities that were in line with this 'procedure-driven figured world' (p. 183). Building on the work of Boaler and Greeno, a useful way of theorising these classroom processes is in terms of the development of learner identities in 'figured worlds' (Holland et al., 1998)—'socially and culturally constructed realm[s] of interpretation in which particular characters and actors are recognised, significance is assigned to certain acts, and particular outcomes are valued over others' (p. 52).

As Boaler (2000a) emphasises, students do not just learn methods or how to carry out a task or to apply algorithms, but they learn 'to be mathematics learners'. Different classroom cultures, constraints, and affordances, provided by different settings and opportunities for engagement in mathematical practices, are likely to influence their perceptions of what it means to learn and do mathematics. Learning how to engage successfully with the mathematics means learning how to identify with the norms of the classroom community. Particular tasks may reinforce practices initiated and propagated by the teacher, or vice versa.

THE STUDY

In a previous study (Pepin, 1997, 1999), I developed an understanding of varying practices in mathematics classrooms in England, France, and Germany using an ethnographic framework. More recently, I investigated, with Linda Haggarty, the ways in which mathematics textbooks are used by teachers in English, French, and German lower secondary mathematics classrooms (Haggarty & Pepin, 2002; Pepin & Haggarty, 2001). This work suggested that the use of curricular materials (such as textbooks), together with the selection of (mathematical) tasks, impacts to a large extent on the mathematical 'diet' offered to students, which in turn is likely to influence students' perception of what mathematics is and what it is to behave mathematically.

For this chapter, I have reanalysed some of the data collected over the years in terms of potential pupil identity formation. The idea of 'figured worlds' as places 'where agents come together to construct joint meanings and activities' (Boaler & Greeno, 2000, p. 173), and which are made up of 'webs of meaning' (Geertz, 1973), was particularly useful here:

> A figured world is peopled by the figures, characters, and types who carry out its tasks and who also have styles of interacting within, distinguishable perspectives on, and orientations toward it. (Holland et al., 1998, p. 51)

The selected data (for this study) consisted of lesson observations and interviews with 10 teachers in each country. I also carried out a task analysis drawing on my knowledge and analysis of textbooks and textbook tasks and their use in classrooms to develop an understanding of how these may link to classroom practices and pupil identity. Textbooks which were originally identified (in 2000) as amongst the ones most frequently purchased for years 7 (6ème, Jahrgang 6), 8 (5ème, Jahrgang 7), and 9 (4ème, Jahrgang 8)[1], and that are still used in classrooms in the three countries, were chosen for reanalysis. The topic of 'directed numbers' was selected for a more detailed analysis in order to exemplify the points made.

I used a procedure involving the analysis of themes similar to that described by Woods (1986) and Burgess (1984). However, it was important to address the potential difficulties with cross-national research, in particular issues related to conceptual equivalence, equivalence of measurement, and linguistic equivalence (Pepin, 2002; Warwick & Osherson, 1973). In order to locate and understand teacher pedagogic practices and classroom cultures in England, France, and Germany, it was useful to draw on knowledge gained from earlier research which highlighted the complex nature of teachers' work and the classroom environments in terms of influences (such as systemic developments and educational traditions) in England, France, and Germany.

The main questions addressed were:

- What kinds of textbook tasks do pupils engage in, and what are the classroom environments in which pupils work/operate in England, France, and Germany?
- How may mathematical tasks, teacher practices, and classroom environments relate to pupil identity construction as learners of mathematics in England, France, and Germany?

THE MATHEMATICS CLASSROOM ENVIRONMENT

Learners of mathematics at the secondary level work in different environments in England, France, and Germany. Whereas in England and France most pupils go to comprehensive schools, in Germany pupils are divided into those going to the local grammar school, *Gymnasium* (about 40%); the technical school, *Realschule*; or the secondary modern school, *Hauptschule*. Furthermore, pupils in the three countries experience different organisations of schooling, which in turn has implications for the 'mathematical diet' they are provided with and experience. In England, most schools apply a 'setting system' to teach mathematics in perceived ability sets, allegedly to raise achievement, despite research (Boaler, 1997; Boaler et al., 2000), providing evidence to the contrary. Different textbooks are used for different sets (and there are different textbooks published for different sets/strands), providing different groups with different mathematical 'diets'. In France, most pupils are taught in mixed-ability groups (and provided with the same textbook)—pupils are said to be 'entitled' to the same curriculum. In Germany, pupils are effectively streamed by the three school types, but within those streams they are taught in mixed-ability groups. The three school types also have their own mathematics curricula and textbooks.

In previous studies (e.g., Pepin, 1999), I identified characteristic 'profiles' of classroom situations in England, France, and Germany. In terms of figured worlds, teachers assigned significance and value to particular practices which are commonly concerned with pupil engagement and assessment of understanding. For example, in the English classroom, the main aim was to (relatively briefly) explain a particular mathematical notion and let pupils get as much practice as possible. Of particular importance was that pupils were attentive during teacher explanations and subsequently worked on their own while teachers attended to individual pupils' needs. The French teachers regarded their main aim as facilitating mathematical thinking by initiating tasks and helping pupils to think around a particular concept, in whole-class conversation, as individuals, or in groups, followed by practice. Thus, of particular importance was that pupils would discover the concept with the help of selected cognitive activities. The main objective in the German mathematics classrooms was to discuss mathematical content. Teachers initiated tasks or discussed exercises from the homework in a conversational style, before giving pupils exercises to practice on their own. They particularly valued that most pupils would be involved in a teacher-led discussion about the mathematical content.

Moreover, there appeared to be particular 'conventions' to which all teachers adhered. For example, teachers in all three countries ask pupils to work on exercises from textbooks for a considerable amount of time so that pupils can practice what has been explained and teachers can monitor understanding. However, in England, many pupils at Key Stage 4[2] and almost all pupils at Key Stage 3 had not been issued with a textbook to use in school and at home; they only worked from textbooks during lessons under teacher guidance. Thus, it is likely that the majority of these pupils only ever had access to the textbook in class and consequently had to rely entirely on teacher guided input. In France, the situation was quite different: Every pupil had a textbook provided by the school to be used in school and at home. In Germany, pupils had to buy their own textbooks which were selected by schools/teachers from a ministry approved range. Thus, already at the outset, there are differences in the roles and importance assigned to textbooks in the figured world of the classroom environment, and for students in terms of access to textbooks.

MATHEMATICAL TASKS

In recent analyses (Pepin & Haggarty, 2007), textbook tasks were analysed with respect to context embeddedness (Freudenthal, 1991). It was found that whereas in English and German textbooks approximately half of all tasks were context embedded, albeit contrived contexts, in French books this was reduced to approximately a third. This means that for any German, French, or English student, at least every second task is situated totally in the abstract world of mathematics, and thus not likely to connect to the world in which students live. As an example of context embeddedness, in the Lambacher–Schweizer 7 (LS7) (Schmid & Weidig, 1994), the reader finds selected exercises that appear to solicit meaning and relate to 'real contexts':

> Explain what is meant by the following:
>
> a- Rita's account says '-85DM'.
> b- The Kaspian Sea has a geographical height of -28m.
> c- The net weight of a box of chocolates differs -15g from the required weight. (Schmid & Weidig, 1994, p. 81, no. 3)

In this case, pupils are expected to link the concept of negative numbers to 'real' situations (such as bank accounts and debt, water level above/below sea level). We found other tasks, in particular in the 'Cinq sur Cinq–6-ème' (Delord, Vinrich, & Bourdais, 1996), where contexts are provided in the introductory activities and are presented as a rationale for 'inventing' negative numbers, but then abandoned in the later parts. It appears that the introductory activities provide a pretence, rather than an actual site for learning, and that the mathematical concepts are the important part and are true with or without context.

In Key Maths 7[2] (Baker et al., 2000), context embeddedness is typically portrayed in the following task, which sets the question in the context of a lift in a building, as suggested in the Key Stage 3 National Strategy for Mathematics (DfES, 2001):

> Here is the control panel in a lift.
>
> | | |
> |---|---|
> | 4 | Fourth floor |
> | 3 | Third floor |
> | 2 | Second floor |
> | 1 | First floor |
> | 0 | Ground floor |
> | −1 | Underground car-park |
>
> a) What number is used for the ground floor?
> b) Where do you go if you press−1 ?
> c) You go from the first floor to the fourth floor. How many floors do you go up?
> d) You go from the third floor to the car-park. How many floors do you go down? (Baker et al., 2000, p. 250)

A similar task was found in the '5 sur 5–6ème':

A tower consists of 12 upper levels and 4 levels below ground. What do you imagine the lift board will look like? (p. 218)

This task was illustrated with a cartoon where a man waits with two heavy suitcases and a backpack in front of the lift, which is 'out of order'.

It appears that in both cases the underlying ideas about negative numbers are subordinated to the context, which largely points pupils toward a context-bound and commonsense solution. Indeed, the whole exercise is no more than a number of context-bound questions which relate neither to each other nor to the underlying mathematical ideas of negative numbers. In KM7[2], there are potentially rich contexts provided that could help pupils to relate to the concepts of negative numbers. Unfortunately, each context is stated in isolation, without explicit connection to the underlying mathematical ideas and procedures and without asking questions or offering tasks that are likely to engage pupils at a deeper level. Questions can be answered using commonsense techniques which are in danger of remaining disconnected and context-dependent. Thus, while appealing at first, the tasks and exercises remain unrelated to mathematical concepts or meaning (M. P. Smith & Stein, 1998); moreover, they seem to obscure the underlying mathematics. It appears that these exercises are simply there to be done (Pepin & Haggarty, 2007).

In summary, it appears that context embeddedness is thought to be important for student understanding of negative numbers, and the contexts used appear to be chosen to relate to situations that students may know. To what extent students actually attach meaning to these contexts, or indeed are familiar with them, is beyond the scope of this investigation, although both are questionable (B. Cooper & Dunne, 2000; Zevenbergen, 2001). How pupils relate to and make sense of these situations mathematically is also less than straightforward. For example, Dowling (1996, 1998) argues that representational differences in mathematics textbooks position learners differently, with those aimed at higher sets encouraging an identification as a mathematician, whereas those aimed at lower sets often obscure the mathematics.

TEACHER USE OF TEXTBOOKS AND MEDIATION OF MATHEMATICAL TASKS

I have argued in the previous section that the nature of textbook tasks conveys a particular picture of what it means to do mathematics and to be 'mathematically fluent' (Kilpatrick, Swafford, & Findell, 2001) in the figured worlds of the mathematics classrooms of the three countries. However, we also need to look at how these textbooks and tasks are used by teachers as a tool or an artefact. The lens of a figured worlds approach enables us to see the importance of how textbook knowledge is mediated by teachers in terms of relationships between learners and mathematics.

Although all teachers emphasised the importance of textbooks in their lessons and for their teaching, there was considerable variation in the classroom use of textbooks. The teachers in English schools all used textbooks regularly, largely for pupils to practice exercises selected by the teacher, following on from teacher explanation of a particular skill or technique. All the teachers were either already heavy users of textbooks or were becoming increasingly so. They reported this increase as being due to lack of time to prepare other resources (Pepin & Haggarty, 2003):

> I use it for the pupils to practise the skills that I've previously introduced, so I expect to teach the topic and whatever method may be necessary and to do all explanations and then I use the exercises in the textbooks for the pupils to complete; ... they don't use it to teach themselves. *We* teach them and the pupils then practise the skills. (Teacher 1, England)

The use of textbooks and the selection of tasks went hand in hand with particular practices that shaped the figured worlds of these classrooms. For example, all teachers in England considered that it would be impossible to use the same textbook with all pupils in a year group; differentiation was seen as necessary to accommodate individual pupils' needs. Instead, they talked about the need for different textbooks for high-, intermediate-, and low-'ability' pupils. They perceived high-ability pupils as needing exercises with interesting and challenging questions and, perhaps, some explanation, whereas they saw intermediate-level pupils as needing plenty of straightforward questions practising particular skills or techniques. The needs of low-ability pupils were heavily influenced by concerns about context, layout, and language demands:

> [for top sets] I'm still looking for practice exercises, but with a bit more challenge to them, so whatever I do, can the textbook exercises take them on a bit further, so that's what I am looking for, lots of practice all the time, with top-set. (Teacher 2, England)

> Intermediate students do not need intriguing exercises otherwise they can't do them! You know, they don't need to be taxed, they've got to be pretty straightforward. (Teacher 3, England)

> [...] but children in the lower sets do not necessarily have a textbook and certainly the Year 11 I have which is a set 9, we do not have a textbook for them, there is not a suitable textbook that we have in the Department for them, which doesn't frighten them because there is just too much in these books, they need to be spoon-fed, they need being given short, concise examples they can follow and then having ten of the same really to do. (Teacher 4, England)

The Role of Textbooks in the 'Figured Worlds' of Classrooms 115

Most English teachers gave relatively brief introductions or rules and wanted a large number of straightforward exercises to practice. They said that pupils needed 'much of the same' to practice.

In line with the German teachers' pedagogic practices, that is relatively lengthy whole-class discussions, *Gymnasium* teachers had a clear perception of where they would use textbooks in their lesson, during the consolidation and exercise phases, but not during the 'acquisition phase' when they would use their pedagogic skills to lead their pupils towards the mathematical concept. Previous research (e.g., Pepin, 1999) showed that *Gymnasium* teachers in particular used a considerable amount of time for whole-class discussion of mathematical ideas or particularly intriguing exercises. Teachers wanted 'difficult' exercises for classwork and 'easy' routine exercises for homework. In addition, mistakes in homework exercises were regarded as a site for deepening pupil understanding. This practice element in relation to textbook exercises was also supported by *Hauptschul* teachers who saw the textbook as an absolute prerequisite for the mathematics lesson:

> [. . .] they have the textbook as a frame and also as a support, and I can give homework from the book, that is important for me. (Teacher 1, Germany; my translation)

French teachers mentioned the three phases of their lessons (activitées–cours–exercises), but they all wanted to do the *cours* (essentials of the lesson) by themselves, without the book, and to choose introductory activities (*activitées*) and exercises (*exercises*) from the textbook. In line with the perceived heterogeneity of their classes, these teachers talked about the need for differentiated exercises:

> What interests me is to have lots of exercises, but at different levels because the classes are more and more heterogeneous, and ... I think of heterogeneity of classes in terms of where I propose distinct exercises according to the level of the pupil ... (Teacher 1, France; my translation)

> [. . .] distinguish between support exercises, more or less difficult exercises, and deepening exercises. (Teacher 2, France; my translation)

To summarise, it can be argued that the mathematics teachers in the study were concerned about different aspects of what textbooks could provide for them, and they used textbook tasks in different ways. It appears that the German teachers discussed the mathematics (to be practiced subsequently in textbook tasks) quite extensively with their pupils, and pupils usually experienced whole-class interaction and 'mediation' of the textbook tasks. French teachers wanted differentiated exercises to help them cope with the perceived different needs in their heterogeneous classes, and German

Gymnasium teachers wished for intriguing exercises to challenge their pupils. *Hauptschul* teachers worried about the literacy demands of particular textbook tasks, in a way which was similar to many English teachers' concerns. However, English teachers also appeared to want a 'scheme off the shelf' to help them plan their lessons and provide differentiated tasks for their different sets of pupils, arguably to make their life 'easier'.

DISCUSSION AND CONCLUSIONS

Connecting tasks in textbooks to students' developing identities as learners of mathematics is not a common or welcome link to make. Textbook tasks are often frowned on, and teachers do not wish to be seen to teach 'according to the book'. However, for better or for worse, textbooks are the main resources used in mathematics classrooms all over the world (Valverde et al., 2002). All of our teachers in England, France, and Germany used tasks and exercises from these books, and in all three countries, pupils are likely to be asked to complete context-embedded tasks.

To engage in the mathematics, pupils must find the task intriguing— something they would like to resolve (Hiebert et al., 1997). This assumes that students relate to tasks in the sense that contexts and situations make them real. However, the majority of textbook tasks analysed were procedural in nature; although they allegedly help pupils to become 'procedurally fluent' (Kilpatrick et al., 2001), they also portray this ritualistic act of knowledge reproduction as necessary to becoming a competent learner, whilst at the same time, it can be argued, obscuring the meaning and concept of the mathematics.

The three countries differed in how mathematics is linked to context and what pupils are asked to do in those tasks. Whereas in the LS7 it appeared that context and mathematical concepts are connected in the tasks, and links are forged between them, in the KM7[2] pupils are asked to do contextualised tasks where contexts are chosen seemingly for their own sake, and with little logical progression or connection to the underpinning mathematical ideas. Most exercises could be done without knowing about concepts of negative numbers: To what extent students may deduce concepts, by simply doing the exercises, is not clear. Interestingly, the French textbook exercises studied appeared to use context as a pretence for introducing the mathematics, a Trojan horse to lead students to the 'essential' section, the *cours*, the mathematical concepts.

To ask what students would learn from these tasks, in terms of the mathematics, needs a closer look. One could argue that, in France and Germany, by addressing the mathematics at the conceptual level, students would gain more insight into the conceptual nature of mathematics, and perhaps its structure, than with English textbook tasks. In order to work 'seriously' on the tasks, texts should provide, and allow students to use, methods

The Role of Textbooks in the 'Figured Worlds' of Classrooms 117

and strategies for problem solving (Hiebert et al., 1997). French textbooks address this explicitly in a separate section ('apprendre a resoudre'), and exercises are organised accordingly. Putting the three countries' textbooks on a continuum, it appears that English textbooks leave it to pupils, or their teachers, to devise or identify strategies to solve the problems in books, and this is likely to be with common sense, whereas French textbooks in particular are explicit about how to solve particular problems.

The message that students may therefore get is that:

- mathematics is simply there to be done (KM7^2), and that contexts and concepts do not necessarily 'talk to each other';
- it is not the contexts that matter ('Cinq sur cinq'), but the underlying mathematical concepts, and that there are strategies to 'reduce' the contextualised problems to 'simple' mathematical tasks;
- concepts and contexts may be connected, and that the formally structured mathematics, including its strategies for solving problems, may be useful in real-life problems (LS7).

Although the mathematical tasks are influential in terms of providing a particular 'mathematical environment' that may exemplify for students what it means to 'behave mathematically', another important factor in the figured world of the classroom is how teachers mediate the textbook tasks and the kinds of relationships with mathematics that they foster. It appears that one of the most important responsibilities for a teacher is to set appropriate tasks, and that these 'set the tone' for particular figured worlds. Teachers in all three countries chose those tasks predominantly from textbooks. What was different was to what extent teachers mediated those tasks and the ways they chose to introduce the mathematical ideas necessary to do the exercises.

Providing students with a series of 'coherent' and appropriate tasks seems important, and these depend to a large extent on teachers' beliefs and the environment in which they work. This was true for the classrooms studied in England, France, and Germany: Within the limits of the system, regardless of whether students were taught in mixed classes (France), setted (England), or streamed (Germany), teachers had some freedom to select tasks that could potentially guide their instruction, and they could mediate those tasks in ways they thought best. These classrooms set the contexts of activity—as figured worlds—and provide the frames where the meanings of actions are mediated and conveyed, affording different potential identities for students. Thus, the classroom environment also needs to be taken into account to see the role of textbooks as artefacts in supporting and shaping, or perhaps 'producing', different identities.

In summary, the lens of the figured world provides a useful tool to study and bring together, as a 'coherent whole', the artefact (textbook and tasks), educational practice (teacher mediation of tasks), and social and structural features of the classroom environment. I have argued that the figured worlds that pupils work in will influence the development of their identities as learners of mathematics. These are linked to the mathematical tasks provided and mediated by teachers, the practices that pupils are engaged in when doing those tasks, and the environment they work in and experience in class—and these are different in the three countries. Whereas one could argue that pupils in all three are 'conditioned' to become 'conformists'—hardly any negotiation about the mathematics and its learning is provided—in England the mathematical diet in textbooks may also encourage learners to become 'common-sensers'. Can one say that in France the 'instrumentalist' identity may be favoured, and in Germany the 'connector', in addition to the 'conformist'? If this link was seen to be strong, one would need to consider to what extent pupils are 'trapped' in these identities, for better or worse, according to what they are offered by their teachers. What kinds of opportunities would need to be provided for change to be possible?

ACKNOWLEDGMENTS

The work and research on mathematics textbooks has been undertaken in collaboration with Linda Haggarty.

NOTES

1. The following textbooks were chosen for reanalysis: Germany: Lambacher–Schweizer (*Gymnasium*)—LS7; Einblicke Mathematik (*Hauptschule*); France: Cinq sur Cinq–5 sur 5–6ème; England: Keymaths–KM7.[2]
2. In England, compulsory schooling is divided into four key stages. The teachers in this study taught pupils in Key Stage 3 (ages 11–14) and Key Stage 4 (ages 14–16).

Part IV
Pedagogy

The three chapters in this part all aim to respond to the question: How do pedagogic practices impact on learner identities in relation to mathematics? In doing so, they have two features in common: first, they explore how existing pedagogic practices, and the discourses they draw on or embody, frame learners' relationships with mathematics; second, they offer insights into how pedagogic practices could be changed or made different in order to facilitate more positive identities amongst learners.

In each of these chapters, the notion of learning relationships, held either between students and teachers or amongst students themselves, stands out as a central theme. For example, in the first chapter, Tamara Bibby (Chapter 11) emphasises the nature of relationships between teachers and students. Using a psychoanalytic lens, she argues that relationships, characterised by issues of emotion and trust, actually frame what is knowable. Here, Tamara presents us with two types of relationship to consider in our understanding of classroom pedagogy. The first is a doer/done to relationship where there is difficulty for each party in acknowledging the other person as separate and where one is required to submit to the control of the other. It is this type of relationship, she argues, which many teachers work to maintain in the classroom. The second is a two-way intersubjective relationship, in which the other person is recognised as separate, is listened to, and is 'held in mind'. It is the latter which the pupils in Tamara's studies appear to seek from their teachers. By way of illustration, she presents the example of Mr. Blake, who is valued by his students because he listens to them and provides space for them to investigate their 'unknown' ideas and explanations without imposing too much control (and risk).

This viewpoint has resonance with the next chapter by Pauline Davis and Julian Williams (Chapter 12). Here we see how one particularly close relationship is played out between two students, Ellen and Craig, in a classroom where sociality is encouraged as part of the pedagogic approach adopted by the teacher. Pauline and Julian argue that the formation of their mathematical identities is afforded by the peer talk which constructs/sustains Ellen and Craig's friendship. Their discourse is striking in the seamless interplay of the two types of talk identified—their highly informal peer talk permeates their sustained engagement with the mathematical activity they are undertaking. Pauline and Julian argue that the students' use of 'hybrid' talk (peer talk used to frame mathematical notions) provides a foundation for accessing mathematics. It seems as if their close relationship aids their use of such hybrid talk and thus provides a pathway into the mathematics they are learning—access which might not have occurred within a different classroom with a different pedagogic approach. Craig and Ellen's friendship has commonality with the close, 'intersubjective' relationships discussed by Tamara, in that their friendship both aids and is reflected in the hybridisation of peer and mathematics talk which makes the unknown (in this case, calculating an estimated mean) safe for them.

However, the authors of all three chapters are also keen to make the point that not just any pedagogic 'intervention' or practice can bring about the kinds of positive learning relationships they have discussed. In recent times, we have seen an increasing desire on the part of policymakers to prescribe certain pedagogic techniques (e.g., pacing, interactive whole-class teaching) which draws on the assumption that providing teachers with a 'teaching toolbox' can and will enhance learning outcomes. It is this kind of rationality and detachment that all three authors seek to deconstruct. This is particularly so in Tamara's chapter, where she argues against this technicalisation of pedagogy, suggesting that many of the practices implemented in mathematics classrooms, such as assessment, pacing, and so on, are placed there to defend against the need for developing 'emotional' relationships with students. For her, pedagogy needs to engage with the emotional labour and risk involved in trying to mutually understand something (and each other), and it must recognise the pain that constitutes not knowing.

This desire for a more holistic view of pedagogy is also expressed by Steve Lerman in his chapter on pedagogic relations and their role in the discursive formation of 'school mathematical identities' (Chapter 13) . Using Bernstein's theory, Steve highlights how pedagogic discourse makes certain subjectivities available to middle-class students by establishing pedagogic relations which are highly 'visible'—and similar to that which they have already experienced in the home. Thus, he argues, middle-class children are more likely to engage with being mathematical in the classroom, whereas this is more difficult for students from working-class backgrounds, who are less likely to recognise what mathematical knowledge should be engaged with in a given situation and are less likely to realise the required 'appropriate' behaviour. Steve presents us with three examples of students engaging in classroom mathematical activities and, like the previous chapters, highlights how mathematical learning is more than just the acquisition of a contained body of knowledge; rather, it is a process of induction into the culture of the mathematics classroom (including its conventions and skills) and dealing with the emotions (such as, pleasure, satisfaction, and frustration) this engenders. He suggests that a focus on identities allows us to recognise this and calls on us to view pedagogy as being about 'the whole person and their becoming rather than just part of them and their knowledge'. There are clear resonances here with both Pauline and Julian's chapter and Tamara's chapter. All three authors suggest that if we wish to adopt a theory of learning which centres around notions of identity, we must recognise that pedagogy cannot be understood in terms of a set of specific practices or techniques which are transferable from one classroom to another. Instead, we need to think about the ways we make certain subjectivities available and accessible to students—in other words, how we bring about different ways of being in relation to mathematics, ways of being which can and do have long-lasting effects.

11 How Do Pedagogic Practices Impact on Learner Identities in Mathematics?
A Psychoanalytically Framed Response

Tamara Bibby

INTRODUCTION

To answer the overarching question on pedagogy, this chapter draws on psychoanalytic theories and research data to explore, from a pupil perspective, the experiences and impacts of different pedagogic practices as they are experienced within relationships. My focus is on the ways relationships and subjectivities frame and constitute what is known and knowable. This focus has arisen from an analysis of two research projects (described later) and a struggle to make sense of the disjunctions and tensions vivid within and across these studies.

Even contemporary definitions of pedagogy fail to account for the ways relationships and subjectivity constitute knowledge. They tend to focus on the teacher's and learner's rational acts and conscious processes. For example:

> [Pedagogy is] any conscious activity by one person designed to enhance learning in another. (Watkins & Mortimore, 1999, p. 16)

I find this kind of definition problematic; its valorisation of conscious processes leaves unconscious desires and resistances unthinkable or abhorrent. But what does it mean to define pedagogy as being about relationships? Relationships are hard work: They involve knowledge and thinking that goes beyond the rational. What space is made for such thinking in schools and classrooms? Where does this leave any notion of a 'pedagogic practice', be that 'teaching styles, tools and resources' or something else? How do the currently highly valued rational 'technologies of teaching' (assessment, planning, inspection, etc.) fit with the idea of pedagogy being about relationships? A psychoanalytic perspective might suggest that they are defences against something; the question is **what** unbearable knowledge have the systems been constructed to defended against (Britzman, 2003)?

Mathematics education has traditionally drawn on cognitive psychologies, such as those of Piaget and Vygotsky or more recent variants (e.g., Engestrom & Cole, 1997; von Glasersfeld, 1990), to theorise classroom learning. Within education generally, and perhaps particularly the hyper-rational world of mathematics, there is a general (although not universal) mistrust of psychoanalytic theories and of thinking about the unconscious; there is often concern about what the two disciplines have to do with each other. Britzman's (2003) suggestion is that both are about making a connection to 'the unknown and the incoherent', the inchoate.

If we focus on curriculum content, the perceived benefits of education, and the systems and institutions, then we may be able to insist (for a time) that education is a rational enterprise: known, knowable, and coherent. But to allow ourselves to explore Britzman's suggestion, we need to acknowledge the ways in which education is about the unknown and the inchoate: the way it is about what learners may or may not know and understand, the way these are inconsistent and shift according to context and time, the importance of feeling valued by a loved teacher, and the way the experiences of education can sear themselves upon our memories as intended knowledge through a teacher's happy use of an apt metaphor or as the burning shame of a moment of humiliation never to be forgotten or forgiven (Bibby, 2002).

As Britzman has noted, even for those motivated to learn between the two disciplines, there is something difficult about the enterprise:

> There is, in educational life, something paradoxical about how the unconscious can actually be considered, particularly because [...] the needs for tidiness and simplicity, so tied to dreams of mastery, prediction, management and control, are all idealizations that defend against the loneliness of institutional life. [... This] institutional ethos of systematicity [...] forecloses any thought of the unconscious and, hence, the work of interpretation itself. (Britzman, 2003, p. 98)

Britzman suggests that, if we could manage to bear not to know, and tolerate the emptiness and loneliness of not knowing, then we could start to learn and to know differently. If we could face the unknown and bear the feelings it arouses, we might begin to be able to stay with the discomfort and think, rather than flee to illusory comfort in unthought. Britzman (2003) suggests this coming into contact with the inchoate is a process which 'entails being able to lose not the object [knowledge] but its idealisation' (p. 163). It is education's valorisation of knowledge and knowing that idealises it: education as the turner-of-keys, the opportunity-creator, the economic driver. Education can do no wrong, it is wholly and commonsensically 'a good thing'. These idealisations of education defend against the terror of the unknown: Can I be successful if I choose not to go to university? What will happen if

How Do Pedagogic Practices Impact on Learner Identities? 125

I don't get Level 4 in my SATs[1]? What would it mean if I was a failure? Or a success?

To foreground the unconscious and explore the less rational aspects of being in school and of coming to know, this chapter draws on psychoanalytic theories to consider how we might develop and understand a learner-derived definition of *pedagogy*. This enterprise raises questions not only about what we might do with such an understanding of pedagogy, but also what it might mean to ignore it.

A PSYCHOANALYTIC FRAMING

The work of Jessica Benjamin (2004) can help us to begin to differentiate various ways of relating. She highlights a distinction between what she calls a one-way, 'doer/done to' relationship and one characterised by two-way 'intersubjectivity'. Most relationships, she suggests, are characterised by a difficulty in acknowledging the other person as a separate centre, someone whose subjectivity is separate from and different than one's own. This is difficult because it requires us to recognise communication as a two-way street. That means we need to accept not only that someone may hear what we are saying, but that it will have effects on them that we can only begin to know if they help us to understand them. To acknowledge the other, to 'hold them in mind', requires that we step beyond ourselves; as far as possible, we surrender our need for control so that we can 'see' and 'hear' what the other person is saying, not what we want to hear them saying. She characterises this as a process of creating a 'third space' (not mine, not yours, but ours), in which, through the construction of 'a diaolgue that is an entity in itself' (Benjamin, 1998, p. xv), both can begin to experience each others' subjectivities and construct an intersubjective relationship.

If this two-way street of intersubjectivity cannot be managed, then we experience communication as a one-way street. In the one-way street of the 'doer/done to' relationship, one party is experienced as requiring the other to submit to their control. Ironically, the other feels exactly the same. An example of this, the tit-for-tat way it oscillates back and forth, and the apparent impossibility of moving on, can be imagined if we think of a 'difficult' class that leaves the teacher feeling 'got at', invisible, and deskilled. The teacher feels persecuted ('done to') by the class, whereas the class simultaneously feel 'done to' by the teacher, who, by their nagging and shouting, appears not to like the class. Each experiences victimisation at the persecuting hands of the other. This kind of relationship is very difficult to move on from. We return to these ideas when we consider the data in the next sections.

The psychotherapist most associated with work on the nature of thought, thinking, and knowledge is perhaps Wilfred Bion, who uses different terms to connect the development of knowledge to a similar relational process.

He talks of the existence or otherwise of a Knowledge Link ('K Link' and its opposite, a 'minus K Link', '–K'). The K Link is that linkage present when one is in the process of getting to know the other in an emotional sense (**learning from** them), and this is to be clearly distinguished from the sort of knowing that means having a piece of knowledge **about** someone or something (Symington & Symington, 2004/1996). For example, a teacher may try to discern whether a pupil knows something by giving them a test; pedagogically, this is seen as an efficient way of measuring knowledge. However, we might recognise that the test is 'easier' (less personally demanding) than risking the emotional engagement that would be required if they were to try to develop a mutual understanding of a learner's knowledge through discussion.

Considered from a pedagogy understood in terms of relational knowing, traditional assessment practices seem problematic. The process of getting to know involves pain, frustration, and loneliness. Obtaining a piece of knowledge **about** a person does not involve these states; rather, it avoids emotional involvement and the pain and difficulty of trying to come to know. Bion stresses that 'there is a continuous decision to be made as to whether to evade pain or to tolerate and thus modify it' (Symington & Symington, 2004/1996, p. 28) and that learning, coming to know, can only happen if the pain is engaged with. This is the difficulty Britzman identified in making a connection with the unknown and the incoherent and that Benjamin located in constructing an intersubjective relationship. As the Symmingtons (2004/1996) explain:

> The [K] Link is a crucial way in which the emotional experience of learning takes place. [If a decision is made not to tolerate the pain of learning it becomes a ...] hatred of learning, [and ...] leads to an attack on the link, resulting in the process being stopped and even reversed. Thus, instead of meaning developing, or thinking being promoted, there occurs a reversal of the process so that any meaningful units become stripped of their meaning. (p. 29)

Forces against the emotional experience of being understood are legion, within both the individual and society.

Traditionally, mathematical curriculum knowledge and process knowledge are treated as 'knowledge about'; my suggestion is that this is highly problematic. What is notable, talking to children of all ages about knowing and learning, is the extent to which they **know** that what and how they know content is intimately bound up in relationships. I therefore suggest that for pupils, curriculum knowledge is bound with the K Links in the classroom: the emotional connections to, and work with, teachers and peers. What is equally interesting is the way many teachers work to deny this link, thereby fostering –K Links'. Perhaps this denial relates to a fear of contact with the unknown or the inchoate, maybe a fear of being overwhelmed or other implications of a more personal intervention being needed? Although, how can teachers themselves learn and develop their teaching if they are not prepared to acknowledge their end of the K Link? If they are not prepared to enter into the difficulty of the **being in and with** in the classroom?

THE RESEARCH CONTEXTS

Two very different research projects are drawn on to develop the arguments in this chapter. Both were conducted in working-class, multi-ethnic, inner-urban schools in London, England. In the first study, Key Stage 3 (KS3) pupils (ages 11–13 years) were interviewed in friendship groups about their learning and teaching. In the second, longer study, researchers followed one class of children for five terms during Key Stage 2 (KS2) from Year 4 (ages 8–9 years) to Year 6 (ages 10–11 years)[2].

Transcripts from the KS3 interviews highlight those characteristics of learning and teaching situations which the pupils most valued. Further, they point up the disjunction between the pupils' thinking and meanings and those of the teacher who interviewed them. Broadly, we can think of this disjunction as between the teacher's insistence on an individual conception of learning and of teaching designed to facilitate this individual 'filling up' with subject knowledge, and the pupils' more collective and social sensibilities and understandings of their learning. Having had a powerful and positive set of experiences with Steven Blake, a maths teacher they clearly valued, the KS3 pupils struggled to make their meaning clear to the (nonmaths) teacher who interviewed them. In direct contrast to the positive and generative learning relationships the KS3 pupils talked about, the KS2 pupils often expressed anger and frustration about their learning and struggled to develop positive relationships with their teachers. In their words, we seem to hear a longing for the kinds of pedagogic experiences that the KS3 pupils described. The tentative redefinition of pedagogy emerging in this chapter is developed from the experiences described by these different groups of children.

A FOCUS ON RELATIONSHIPS?

The aim of the KS3 interviews was to explore the connections children had experienced and made across areas of the secondary school curriculum, as well as the kinds of learning and teaching they enjoyed and the contexts in which they experienced these. While analysing the interview transcripts, I was struck by two things: the love that students felt for Steven, their maths teacher; and the teacher/interviewer's and pupils' different agendas.

In response to the teacher/interviewer's question about whether 'the way you learn in maths lessons [is] different to the way you learn in other subjects?', pupils struggled to describe the ways in which the learning and teaching was different in different curriculum contexts. The root of their difficulty was not particularly that this was an unexpected question, but related rather to the different focus of the teacher and the pupils and the miscommunication under which they were labouring. After a long group-constructed description of the practices of one particular teacher, the teacher/interviewer asked:

> *Teacher/Interviewer:* Is that different than other subjects?
> *Pupil:* I think it's different teachers
> *Teacher/Interviewer:* No, we're talking about subjects.

Throughout the KS3 interviews, the teacher/interviewer was thwarted in her desire to focus on curriculum subjects. She was not able stay with the inter-subjectivity the pupils were focussed on, and she kept trying to move back to an abstracted notion of a mathematics that could exist outside relationships. This may have related to some professional discomfort around talking to pupils about colleagues. Or it may have been that, in common with many people, she had so completely split the maths from the people doing the knowing that she had lost any sense of knowledge being generated and held inside relationships. Certainly the thrust and direction of all formal educative processes would support her in developing such a split. A suggestion that flows from this seems to be that, for secondary school pupils at least, what a subject is, epistemologically speaking, is as much about the teacher and the room as it is about the content or other more adult concerns.

What does a focus on relationships look like? I suggest that there were two themes which characterised the KS3 pupils' talk and that these can also be found in the KS2 data. These overarching concerns are: (a) mutuality, and (b) being seen and valued; it is the second of these concerns that is the main focus of this chapter. A number of smaller themes nest within these. For example, the importance of a teacher's perceived generosity and caring might be understood as gestures that confirm the value a teacher places on the group or an individual:

> *Pupil:* He gives you like activities to do instead of just going straight on working in books or something he gives you more activities to do and if you're in science and stuff they just make you look in the board and do other things in books and stuff.

Here (and elsewhere), the students seem to be indicating that their teacher 'gives them stuff', that there is interaction, a movement from the teacher toward them as people. They feel thought about, and they seem to be valuing this and expressing gratitude and pleasure at what they experience as his generosity in providing social and emotional space: a generosity of spirit perhaps. Such generosity and caring were identified as being particularly associated with giving time: the planning of special activities, the careful marking of work, the playing of a game, and a comment such as 'I was thinking about what happened in the last lesson'. We might think of all of these as examples of an individual or group being 'held in mind' (more on this later).

MUTUALITY

When asked how they learn in maths lessons, some KS3 pupils mentioned what we might expect: equipment, working things out, and memorising, but others talked slightly differently about mutuality and the 'groupness' of the class, and particularly the larger notion of the class with teacher: 'We cooperate with each other and the teacher'. Throughout both projects, and especially the KS2 project, this child awareness of and focus on the group is striking and stands in marked contrast to teachers' focus on individuals. I briefly mention these differing emphases on the importance of the group and the individual because it would seem to lie behind much of the confusion that exists between teachers and pupils. The individualism inherent in the opening definition of pedagogy might be unwelcome to many learners.

BEING SEEN AND VALUED

Experiencing a two-way, intersubjective relationship, being 'held in mind', requires one to be 'seen' and 'heard'. That is, to be accepted intersubjectively, one's subjectivity needs to be available to and accepted by the person at the other end of the two-way street. Understanding this takes us some way toward making sense of the value children from both schools placed on such experiences.

One group of KS3 pupils spent considerable time and effort trying to explain the importance of their pedagogic relationships by contrasting their experience of being with Steven Blake, their usual maths teacher, with their experience of being with a student on teaching practice. Here we begin to get some idea of the importance of being 'seen' for what and who you are, individually and as a group. The relationship with Mr. Blake might be characterised as being experienced as 'intersubjective', whereas the experience with the student teacher has 'doer/done to' characteristics. Later, I further underline the desirability of an intersubjective relationship by considering an interview with a group of Year 5 girls.

> *Teacher/Interviewer:* What is the best way of learning in maths lessons?
> *Pupil:* Doing it off the board and the teacher helps you out [a reference to Mr. Blake] cos when you do it from the books [the student teacher] just speaks and gets annoyed.

> *Pupil:* If you get to speak in front of the whole class explaining why you think it's this and like everyone sharing their own methods [reference to Mr. Blake], instead of the [student] we've got right now [. . . who] says we can only use this method. [. . .]
>
> *Pupil:* Sometimes we ask her like for help and she doesn't understand it. She tells us what she thinks I said and then walks away and then helps like five other people before she comes back.

In these comments, we hear pupils commenting on the student teacher's desire for control and that she is not listening to them: She gives them what she thinks they need without asking and hears what she wants to hear, not what is said. In this sense, the relationship they are complaining about can be characterised as being 'doer/done to', and within such a relationship one cannot be held in mind and remains unseen and unheard. The contrast here is between a teacher who is currently unable to listen and one who can, and who can also provide the psychic and emotional space for them to explore explanations and ideas without needing to exert a particularly stifling kind of control.

BEING NEITHER SEEN NOR VALUED

The KS2 project was planned to give space for children's voices and interests. In it we get some insight into the fury of children who have a teacher (Rachel South) who might be thought of as Steven's antithesis; a teacher who had bought into new managerialist discourses (S. J. Ball, 2003) and who believed (at some level) that children learn because she tells them to learn. The lack of psychic and emotional space in this classroom was startling. Here, Sabrina explained what it was like being in their classroom. This comment and the following two come from the same group interview:

> *Sabrina:* OK, Miss South, she sometimes ignores me and stuff [. . .] It hurts my feelings. [This makes it] harder [to learn] and like I open my book and I see that I get all the questions wrong, wrong, wrong, wrong [. . .] and it's not my fault that I got everything wrong, it's Miss South's fault that I got everything wrong. [Then] I feel guilty [because...] Miss South ignored me [. . .] I don't know why I feel guilty [. . .] it's still quite my fault because I got the questions wrong a bit, but it's normally Miss South's fault.

The experience of Miss South who learned **about them** only through their work was experienced as diminishing in some way. That she didn't value them enough to **learn from** them left difficult feelings of blame and

How Do Pedagogic Practices Impact on Learner Identities? 131

guilt—feelings we might associate with some nebulous notion of persecution (of being 'done to').

Feeling 'done to' can result in a chronic negation of self. Sabrina's experience of guilt seems linked to an idea that there was something that she had failed to do. It is as if she felt personally responsible for her own invisibility, although she struggled with this feeling. Could it really be her fault that she was unworthy of Miss South's love and notice? Or was it Miss South's fault too? Sally seemed clearer and less willing to struggle with complex feelings of complicity and doubt. To her it was clear that Miss South was unfair because she didn't like her.

> *Sally:* Well I feel like that, it's because Miss South like, I put my hand up and she never chooses me, especially like in maths, she loves the other year 5 class and then like um she blames me if I've got it wrong, it's like 'Sally you don't understand' but it's her, she doesn't understand and then when I'm ignored I don't like it, I feel left out [. . .] But then she says 'I only ignore you, it's because you're so clever', but then that's not true. [. . .] I think it's just she doesn't like me [. . .] When you keep on being left out, you think of something, you think that oh maybe they don't like you and that's how I feel. [... So] it's a bit difficult to like concentrate and then she's like 'you're not concentrating properly', but then when you tell her that 'you're leaving me out' then she doesn't know how you feel because it's not happening to her! [. . .] Cos she's being, like everyone's surrounding her going, 'Miss South, Miss South!'

Sally struggled to know how Miss South could know and understand the loneliness of not knowing, of being a learner. How could she possibly empathise? It wasn't happening to her. For her, Miss South can experience the precious safety and value of being held in mind, whereas Sally cannot; she expresses envy toward the teacher, who is believed to be seen and thought about by everyone in the class. Feelings of envy are very difficult to process, and here we see them turning to anger and violent fantasies:

> *Sabrina:* If I was Miss South yeah and Miss South was me yeah, I'd just squash her like a fly.

So, unlike Mr. Blake, who was talked about with respect and gratitude, Miss South was engendering fury. Over the course of the year, the children in her class repeatedly tried to form relationships with her, and she repeatedly refused to recognise or respond to their attempts, holding herself apart and aloof. The only relationship possible with this adult was one of subservience to her domination. Again, the themes are a demand for empathy, for listening, for being 'seen', but here we see the effect of these not being met, of the children feeling unseen, unempathised with, treated dismissively, and devalued—of a teacher (and school) whose parsimony with time and effort was experienced as extreme, ungenerous, and unthoughtful (although the story constructed by the teachers was of economy of effort, not reinventing the wheel, etc). The teacher–pupil relationships in this classroom might be characterised as 'doer/done to' and as demonstrating a destructive –K Link in which (mathematical and self) knowledge was actively destroyed.

MATHEMATICS LEARNING IN THE PRESENCE OF A –K LINK

So what does this mean in terms of the way the primary school children came to relate to the mathematics they were learning? This is difficult to answer; it is hard to tease apart the mathematics learning from other learning in the classroom. However, as in so many schools and classrooms, mathematics lessons did provide an extreme environment.

Time, the curriculum, the assessment procedures, and the 'vital necessity' of success in mathematics are some of the institutional systems that we have put in place to defend against the unbearable knowledge that we have to develop relationships to learn and to enable others to learn. Indeed, throughout the KS2 data, there are many familiar stories of a focus on speed, right answers, and how things are done and recorded, rather than what is done and recorded and why.

MUTUALITY REVISITED

The desire to both belong to a group and to remain a visible individual holds a tension for all of us (Bion, 2004/1961), yet in today's educational climate, the pressure to be an individual is nearly overwhelming. For children in both projects, being part of a group and the sociality of learning and knowing were paramount, although their teachers and the school system drove them apart. Focussing on the relationships formed with the teachers and the maths, we can see that these are at the heart of the possibilities we have for knowing ourselves as learners. During a group interview, Muhi and his friend Matthew struggled to make sense of why they had both been moved from the 'middle' maths group: Muhi into the 'bottom' group and Matthew into the 'top' group. They explained first that they'd been working together and thought they were getting the same marks in their work and tests. Unsurprisingly, Muhi is particularly indignant and struggles most:

> *Muhi:* Yeah, I got [moved] to the lowest [group]. And Miss Middleton hates maths, so guess what she does? She says 'right, get all your

times tables done' and then she gets paper and we just have to colour, like reception.
Matthew: Miss Middleton is fun. I wish I was in her maths group.
Muhi: But she hates maths. She doesn't even learn us maths. It's [very] boring, we just had to colour like reception [. . .]
Emran: When Miss Middleton took us just for a bit she never taught us it and I think it was just a waste, I don't know, of teaching. Because I learned more from when we had to do sticking to make a collage.

Matthew seems to try to make it okay for Muhi to be in Miss Middleton's bottom maths group, it must be fun. But for Muhi, there is only frustration; he is left not learning a subject he used to enjoy, and he feels diminished by the experience. Eventually, Emran confirms his interpretation of the lessons: Little learning happens. Muhi feels helpless and angry faced with a teacher who refuses to acknowledge or form a two-way relationship with Muhi-who-enjoys-maths and who assumes that all the children in her bottom group share her wish to avoid the subject. How can he know how to behave with this teacher? Here, in the face of a –K Link, his understanding of himself as a collaborative and successful mathematics learner has been stripped of meaning, and he is left empty. His energies, once focussed on learning mathematics, shift as he struggles to understand his sense of loss, disappointment, and bewilderment: Who am I now that I have been demoted to this nonmaths maths class?

PEDAGOGY RETHOUGHT: SOME INCOMPLETE THOUGHTS

Education in England today is fixated on idealised imagined futures brimming with fulfilment gained through knowledge and the educative enterprise. The pressures of accountabilities (the 'market' of parental choice, surveillance, league tables, etc.) keep this idealisation active. For teachers who, like children and the rest of us, want to feel seen and valued, who desire the experience of being held in mind in an intersubjective space, their accommodation to the pressures they experience may seem sensible, perhaps even inevitable. But we have begun to see how these accommodations can be experienced by the children in their classes and the value children place on those teachers who can resist passing the pressures down.

Idealisation is impossible to live with; putting anything or anyone on a pedestal is a recipe for disaster. Through their continuing connection to the sociality and relationships of the classroom, the children had some awareness of the complexity inherent in both wanting to belong to the group and to be special for themselves, of wanting perfection and resenting the envy it engenders, and of the difficulty of accepting the uncertainty that contact with the unknown and the incoherent bring.

In the classroom characterised by one-way relationships, the–K Links that prevailed stopped children like Muhi from making contact with their inchoate mathematical and social knowledges: Knowledge was thereby actively lost. The difficulty of wanting to be seen and valued while feeling invisible and worthless diverted attention away from academic learning and toward making sense of the intensely painful feelings that took their place. Comments in the field notes record the agony of this:

> Rhatul is still working out his problems but to confirm his progress he calls out: 'Miss, I'm on number thirteen'! I am intrigued by this constant updating and ask him why he calls out to the teacher. He looks a bit shy but then answers quietly 'because sometimes she doesn't care what number I'm on'.

And in an interview:

> *Minnie:* I always have my hand up and Miss South never picks me, like I had my hand up before Sally and Miss South picked Sally.
> *Researcher:* That does sound difficult, how does it make you feel?
> *Minnie:* Broken.

Throughout this, the KS2 children continued to think about the teachers in ways that their teachers appeared not to reciprocate. The children thought and talked about why the teachers behaved as they did. They spent a great deal of time and energy trying to understand and lessen the pain the teachers unwittingly generated. In an act of empathy, one group of girls explained to the researcher that they believed the teachers did not think about them because they were stressed; simultaneously, they bore witness to some damaging effects of idealisation:

> *Researcher:* Why do you think the teachers get so stressed?
> *Minnie:* Because they want it to be perfect.
> *Rani:* Yes, it all has to be perfect.
> *Minnie:* They think we are perfect child, but we are not. We are just children.
> *Rani:* Yeah, every child ain't perfect. There is always something...
> *Minnie:* Wrong with them.
> *Rani:* Not wrong with them. But they ask that we all be perfect.

When we lose our idealisations and when we stop defending against the difficulty of knowing and **learning from,** we are forced to look at our difficult feelings. It seems that sometimes children are more willing to do this than we are as adults. Sabrina's struggle with guilt, blame, and complicity begin to show what is at stake here. Idealised, I can never be seen or heard and I can never experience being held in mind (because I cannot truly exist); I can only ever feel persecuted. In the unsatisfactory and defensive tit-for-tat, attack, and counterattack of a 'doer/done to' relationship, if I feel persecuted by you, you will feel persecuted by me.

If we accept that a definition of pedagogy needs to be founded in relationships or relationality, then we cannot continue to operate with the idealisations of mathematics and of learning and knowledge that pervade schools and maths classrooms. Intersubjective relationships cannot survive idealisation. In this chapter, I have tried to show that doer/done to relationships result in knowledge being lost, in the energy for learning being diverted. Paradoxically, giving time and space to the development of intersubjective relationships may create more generative 'academic' learning time. To begin to address this notion requires a radical rethink in mathematics education. This is not a new idea, although I may have cast it in somewhat different terms. The inchoate and learner-derived definition of pedagogy I have begun to articulate here leaves any decision to ignore its premise highly problematic and deeply unethical.

NOTES

1. In England and Wales, students are assessed against the levels of the National Curriculum in SAT tests.
2. ESRC project number RES-000–22–1272 'Children's learner identities in mathematics at Key Stage 2'.

12 Hybridity of Maths and Peer Talk
Crazy Maths

Pauline Davis and Julian Williams

In this chapter, we show how students' collaboration regarding their work in class can sometimes facilitate a sociable 'community-influenced' classroom mathematics talk. The social affordances of collaboration, and indeed mathematics as a human, social activity, have been examined by many others, although sometimes under the guise of equitable mathematics or inclusive mathematics pedagogy (e.g., Boaler, 2000a; Boaler & Greeno, 2000; Nasir & Cobb, 2002) and earlier in the area of 'cooperative' classrooms. Such research has consistently pointed to the value of encouraging student–student interaction as a means to foster more positive identifications with mathematics, often for those who might be described as struggling learners (e.g., Boaler, 2000a).

This chapter provides an example of how classroom talk is fundamentally social and how students' sociality can be harnessed in ways to engender a living, more vibrant, and even synergetic kind of mathematics talk (Yackel, Cobb, & Wood, 1999). We explore an example of students' everyday classroom talk, where students are studying for an optional advanced qualification in mathematics usually taken at age 17. The classroom is within a college in a provincial British town, which draws predominantly from a local White working-class community. We focus on an example of how an alternative 'street-talk inflexed' term for estimated mean comes into being during an episode when two students explore together their alternative solutions, as they search for their error.

Here we draw parallels with researchers in the field of literacy (e.g., Hicks, 2002; McIntyre, Roseberry, & Gonzalez, 2001; Moll & Gonzalez, 1994) who show how allowing 'community' discourse into legitimate classroom talk can provide for hybridity (Bakhtin, 1981). We suggest that captured within the students' renaming of the 'estimated mean' as 'a crazy 20' is a hybridisation of 'peer' talk with 'maths' talk.

We speculate that aspects of the classroom pedagogy allow this new vibrant mathematics talk to emerge, in contrast with classrooms, where teacher talk dominates, where social talk becomes separated from the maths talk, and where peer talk is muted, hidden, or even oppositional. We suggest this may indicate that a blurring in the regulation/classification and framing (Bernstein, 1990) of the traditional pedagogic codes may both foster a more vibrant kind of classroom mathematics talk and more sociable ways for students to be mathematics learners. We suggest that this may create a synergy with some learners' social identity so that they may identify with mathematics. Here, identity is viewed as a 'nexus [... of] multiple forms of membership through a process of reconciliation across boundaries of practice' (Wenger, 1998, p. 163). This is a process of the 'reconciliation' of multimembership, where the individual comes to an understanding of themselves across boundaries. The notion of reconciliation is helpful because it highlights the relations between different forms of multimembership—for example, of a learner with their peers outside school 'on the street' (without the purview of adult or parental restraint) and with their classmates in the classroom.

BACKGROUND

The mathematics classroom we draw on here is part of a sixth form college (a postcompulsory education college with students mainly between the ages of 16 and 19) situated in a predominantly White working-class city in the former industrial North East of England. Almost all the students in the class live in the locality of the college, and they typically come from areas that can be classed as being of modest means based on the ACORN demogeographic categorisation (http://www.caci.co.uk/acorn/whatis.asp). The students in the mathematics class range from 16 to 18 years of age and typically had relatively low entry qualifications in mathematics. This is part reflects a deliberate 'open' access college policy on recruitment.

The featured classroom is taught by an experienced teacher whose practice is underpinned by deeply held beliefs in constructivist pedagogic principles (Wake et al., 2007). Based on the results of a self-assessment instrument measuring teacher centricity (Pampaka, Williams, Davis, & Wake, 2008), this teacher was the most student-centred of the sample. For example:

> *Teacher:* There's a sense that I've achieved the purpose [. . .] I've found out what they've come with and what they haven't come with so ... we can work with that now. [. . .] I want to get students to think about the maths, I want students to understand, I want students to connect ideas together, to see all those things that go together and I don't think a text book did that.

Her lessons tend to involve a mixture of collaborative group work and whole-class teaching, which are used in the context of an established collaborative and inclusive classroom. She herself emphasises that social relations in the classroom are vital for many of these particular learners, and she employs specific strategies to improve their communication with each other as well as with her. In interviews, many of the learners remarked on this—how in this maths classroom, everyone knows everyone else (in contrast to some other subjects, such as sociology). Many also refer to maths as 'fun'. One student, for instance, told us that the main reason she had not dropped out of college was because of the social life and the friends in her maths class.

The particular lesson that is used in this chapter is the second in the statistics unit of the programme. The students have been asked to calculate an estimated mean average for data that has been grouped by the students in different ways (so not all answers are exactly the same, but they should be approximately so).

TABLE TALK

This section provides extracts of classroom talk that we use to illustrate the argument being made. The first two extracts serve to demonstrate peer talk between two students Ellen (age 17) and Craig (age 18). In the third extract, their focus turns to the problem they are solving. The remaining three extracts demonstrate the argument presented here with regard to hybridity. In this first extract, Ellen and Craig are keen to spot an error in one of their workings. They are sitting at a table in a classroom set up cabaret style, and the table has a microphone discreetly placed on it. The episode selected takes place an hour or so into the lesson:

> *Craig:* Please do not destroy my calculator ... I pinched this.
> *Ellen:* It's annoying.
> *Craig:* No, it's just because it's not like your calculator.
> *Ellen:* It's not like my calculator.
> *Craig:* It's not like your cheating, 'I'm going to cheat in exams' calculator.
> *Ellen:* My calculator is amazing. Never diss the calculator.
> *Craig:* I'll diss what I want.

In many ways, this talk is quite unremarkable and could be seen simply as two students' engagement in some fairly innocuous off-task banter while they work on the task. What we see here, however, is open, sanctioned talk, in that the teacher makes no attempt to suppress their conversation because they have some freedom to be social within the context of the lesson. The teacher did not necessarily know whether this was going on; the point is that the teacher trusted the students and did not need to know exactly what was going on and, by so doing, sanctioned the talk.

We see their interaction regarding the calculators as serving to maintain their relationship sociality as they inspect each others' mathematical workings to jointly find where the discrepancy in their responses to the problem stems from. Ellen is frustrated with Craig's calculator because she can't work it sufficiently well, so she hits it. Craig light-heartedly takes control by asserting that the calculator is his possession and she should be careful with it (Line 1). He implies that Ellen is annoyed with his machine because she does not have the know-how to use it properly because it is not like hers. Ellen reverses the position when agreeing that Craig's calculator is indeed not like hers, which she implies is a much better model. Craig maintains the thread, but by calling hers a 'cheating calculator', both recognising that she does have the better model, whilst attributing his with integrity and arguably himself with the know-how.

Next there is an abrupt change in voice as Craig sings half under his breadth to the tune of the popular song *Camptown Races*, and Ellen responds to his childish slapstick with a matter of fact agreement, which swiftly brings this aside to an end as they continue with their problem solving:

> *Craig:* Oh, Ellen smells, Ellen smells, Ellen is a smelly pants, doo-dah, doo-dah. Ellen smells of pooh! Ellen is a smelly pooh pants.
> *Ellen:* I do.

They continue their social engagement through their voluntary efforts to pinpoint a discrepancy in their answers to a particular problem, and their talk becomes less colourful for a while.

> *Craig:* I'm angry now [said ironically as Ellen is chewing his pencil].
> *Ellen:* Did I just do divide? No, I didn't.
> *Craig:* You just did times.
> *Ellen:* I've forgotten how much I like pencil.

For a short time, two strands of talk that is maths talk and peer talk seem to develop simultaneously, although the pencil conversation is soon dropped. We also noted such interweaving of social ancillary conversations into the main thread, which is to do with the maths problem when later Ellen picks up again the 'poo talk' thread by scribbling a note that Craig smells.

> *Craig:* Yeah it is; it's a space to write them. Oh, I'm allowed it [the calculator] now, just as I've actually got to work it out. [He means he has just got it worked out.]
> *Ellen:* Add it up and divide by seven. Do you divide by seven? [There are seven intervals, each of which has a frequency, so the divisor is the sum of the seven frequencies.]
> *Craig:* Noooooo, you divide by the total number. I bet that says something weird.
> *Ellen:* Is that a 13 or a 15? I don't know.
> *Craig:* [Calls over the teacher.]

> *Ellen:* Do you get 119? You what? [Ellen realises Craig's answer of 20 is way off hers of 119]
> *Craig:* Great, I've **got my crazy thingy there, estimated mean**, how?

This is the first time Craig connects crazy with the estimated mean. He calls the 'thingy' (estimated mean) crazy, and his use of the pronoun 'my' indicates that he takes ownership of his different estimate. Earlier in the lesson, Craig experienced difficulties with similar problem solving, so we speculate that maybe he is beginning to suspect that it is he who has made the error. Craig calls the teacher over for help.

> *Teacher:* You've got that? Crumbs, [She confirms that his answer is way off.]
> *Ellen:* I added all them there and then divided by total frequency ... got that.
> *Teacher:* Yeah it makes more sense, doesn't it?
> *Ellen:* Yeay! [Pleased with herself—we note the friendly rivalry between the pair.]
> *Teacher:* Talk me through what you've done. [Directed to Craig.]
> *Craig:* I added all them together, and then divided by the total frequency, and that's what I got.
> *Teacher:* Where's 25 times 6? Talk me through where that comes from.
> *Craig:* It's the midpoint of that group.
> *Teacher:* Of which group?
> *Craig:* That group.
> *Teacher:* 100 to 150?
> *Craig:* Yeah.
> *Teacher:* What's halfway between 100 and 150, Craig?
> *Craig:* 25. 125. [Craig readjusts his response to 125 as his arithmetic error dawns.]
> *Teacher:* 100, yeah, there you are, you've lost all your 100s.

Craig's error has been identified by the teacher. It was an arithmetic slip, where he had mistaken the midpoint of 100 and 150 as 25 instead of 125. This slip resulted in his final estimate of 20, which was unlike the estimates from other students, which hovered around 120.

> *Teacher:* Suzanne, have you finished?
> *[Ellen and Craig to themselves at the table.]*
> *Ellen:* How'd you get that?
> *Craig:* I don't know!
> *Teacher:* Could you just add those up for me? [The teacher has moved to another table.]
> *Craig:* Crazy20. [Directed toward Ellen.]

Here we see Craig has become more specific, and his '**crazy thingy estimated mean**' becomes '**crazy20**'. We note, however, that the teacher didn't pick up on Craig's expression, so it went unnoticed within the classroom as a whole. In this sense, his 'crazy20' didn't make it beyond his table into general classroom discourse.

COMMUNITY DISCOURSE CROSSING THE BOUNDARY INTO CLASSROOM MATHEMATICS TALK

We see how peer discourse is used as they talk, for example, in Ellen's use of the term 'diss', a short hand for 'disrespect'; arguably, there is a connection here with African-American peer talk, which is now widely used in some British communities. The 'intertextuality' within Ellen's discourse is typical of talk in very many conversations and situations. It is precisely because such talk is typical outside the classroom in all kinds of circumstances that it is of interest here, and indeed there are many classrooms where students draw on a range of discourses. The interest in this chapter is in exploring why such talk may offer a potential for change which renders it of significance here. We note also how Ellen and Craig (both White British) use the terms 'crazy' and 'weird' in relation to aspects of their mathematising.

For example:

Craig: The only thing you need to know about electrons is that they've got a negative charge and they do **crazy** stuff and they spin in directions.
Ellen: Yeah, **crazy**.

And in relation to 'histograms':

Ellen: Is one of them like the … or something **crazy**, but done into a bar graph?
Ellen: You've got to do something **crazy** with them.
Ellen: Yeah, these boxes are **weird** [that is, of different widths].

The exception to this is Ellen's recourse to its standard usage when she says, 'Is that because **you're crazy**?'

We suggest that 'crazy' is being used as a 'filler' and in a sense denotes a black boxed procedure—any procedure, that may be highly important, which is not known or understood by the user. We do not have enough instances of the term 'weird' to situate it in this way, but both 'weird' and 'crazy' can be seen as imports from 'peer talk'.

When Craig uses the term 'crazy' in 'Great, I've **got my crazy thingy there, estimated mean**, how?', we have an example of hybridity—both 'crazy thingy' and 'estimated mean' are used in conjunction with each other when

Craig draws on both his peer and mathematics discourse. Similarly, when Ellen refers to the essential distinguishing feature of histograms (compared with bar charts) as 'weird', she is using her genre of (or register for) peer talk to mark out and make the mathematics her own.

Later, when Craig refers to his '**crazy 20**', he drops entirely any reference to the mathematical terminology of an 'estimated mean', which has now become implicit given the discursive context. It is a hybrid construction—'crazy' combined with '20'. That is, 'it is an utterance which belongs, by its grammatical and compositional markers, to a single speaker, but actually contains mixed within it two utterances, two speech manners, two styles [. . .] two semantic and axiological belief systems' (Bakhtin, 1981, p. 304). Such hybridity in Craig's language use is important because it opens up possibilities for advancement or regrowth because the talk is in a sense at the boundary or edge of two or more discourses (or systems).

Craig's use of 'crazy 20' might best be described as an instantiation of organic hybridity because there is a combination and fusion within it, 'but in such situations the mixture remains mute and opaque, never making use of the conscious contrasts and oppositions' (Bakhtin, 1981, p. 360). This 'unconsciousness' might suggest a danger for the future—at some point, Craig may need to become clear about the inappropriateness of such peer talk in some situations. Organic hybrids can be 'profoundly productive historically; they are pregnant with potential for new world views' (Bakhtin, 1981, p. 360). Each generation makes the culture (including mathematics) in a new form, and it can be argued that organic hybridity describes the historical evolution of all cultures (Lo, 2000). Despite having transformative potential, organic hybridity does not necessarily disrupt the sense of order and continuity as long as the keepers of the culture accept it as his teacher does.

The term 'crazy 20' then is a construction personal to Craig and Ellen, which captures the particular context of Craig's calculation of an estimated mean as they worked jointly on the task of locating his error. The crucial thing is that the use of 'crazy' from peer culture has become part of the way they talk about maths itself; it has become constitutive of maths talk for them, and, hence, we claim the description as hybrid is justified. As such, this goes one step further than ancillary talk, when the peer talk is accepted but kept separate from the maths in the classroom. However, we also note that there was a mixture of ancillary talk and talk that is constitutive of maths at their table. So what then is the significance of 'crazy 20' when it comes to identification with mathematics?

MATHEMATICAL CLASSROOM PEDAGOGIC CULTURE AND IDENTITY

In some ways, the pedagogic culture of Ellen and Craig's classroom is fairly atypical of our observations across sixth form colleges and further education institutions, in that it was underpinned by the teacher's principles and practices (that we might describe as 'constructivist'). However, the episode we draw on is simply that of the two students working together on individual work, which becomes a joint activity when their task becomes to spot their error.

In one sense, we are stating the obvious when we point out that when the students were talking together they were being social. We argue, however, that the open acceptance of this sociality within classroom culture is crucial to the ways in which students will identify with mathematics. Indeed, the literature suggests that sociality might be especially important for those students whose community cultures are not well aligned with the dominant discourse of school as an institution. Hicks (2002), for example, concludes that because 'the forms of action and knowledge that students embrace are strongly tied to the identities that emerge from family and community contexts, conflict can arise between an institutional system aligned with middle-class practices and the life world of working-class students in particular' (p. 2). Kendrick and McKay (2004) argue in a similar vein that when school practices do not afford spaces for belonging and when students are unable to place the cherished identities they live at home and outside of school in dialogue with new identities they encounter at school, they turn away to other values and practices as points of identification and connection.

Thus, we argue that Craig and Ellen's banter about calculators and Craig's messing about with the 'smelly poo' talk is important when it is significantly intertwined with mathematics. As we saw in the earlier extract, doing maths becomes part of that identity work. For instance, Craig in relation to this episode later reveals in conversation about his arithmetic error, 'but [later] I redeemed myself'. Indeed, at one point toward the end of the lesson, in banter with Ellen, he says, 'I am a man'. Both examples highlight how gendered identity work is being performed.

We note also that in this classroom, opportunities for students to draw on significant elements of their identity often go beyond their peers to include the teacher. In the following extract, we show how Craig was keen to engage his teacher in a similar kind of banter to the calculator talk:

> *Teacher:* Hang on. So what will that one? Was that 175, that one? I told you, you should have a nice table, didn't I? Who was right?
> *Craig:* Oh God [meaning don't gloat over me now].
> *Teacher:* I'm glad you record this. [Directed to the researchers] Teacher's always right! Oh dear.
> *Craig:* You're not always right. I will prove you wrong one day. I've proved you wrong before!
> *Teacher:* I am this time. Just this once I am.
> *Craig:* I won before, on the first day. Still remember that bloody day. I got a Mars bar for that.

We note it is important that the teacher validates their ancillary talk by joining in, albeit she doesn't adopt the same language as they—she can be 'cool' but never truly their peer—but we see that she accepts and understands it.

If we look at Ellen and Craig's classroom table talk (across the extracts), it strikes us that, throughout the episode, the peer-to-peer tenor (in the terms of Halliday & Hasan, 1985) of their conversation broadly speaking stayed the same, signifying that the relations between the two of them stay constant even when the topic of the talk changes. The notion of tenor then is important because we see the mathematics 'field' has become entwined with that peer tenor. (For a fuller discussion of field and tenor, see J. Williams, Davis, & Black, 2007.) In other words, they talk about mathematics using broadly the same tenor as when they talk about chewing pencils, albeit we would not expect the talk emerging as public to be in connection with other aspects of their lives that they hold significant—connections that we claim are necessary if mathematics is to take on significance in their lives (see Cobb & Hodge, 2002).

This is why sociable mathematics may be more appealing for those students whose neighbourhood discourses are in opposition to those dominating in schools or colleges. When peer talk is not encouraged or allowed within the classroom, students may experience proportionally fewer opportunities for connection-making than those students whose neighbourhood discourses are better aligned (see Davis, 2007a, 2007b). This also explains why what constitutes effective pedagogy may vary with social class (Borich, 1996), ethnicity, gender, and other social differences. Moreover, there is a growing literature stemming from research in school effectiveness (e.g., Lauder, Robinson, & Thrupp, 2002; Thrupp, 1997, 1998) that looks at the impact of what is sometimes known as a 'school-mix' effect, which considers the impact of the sociocultural make-up of the classroom on school processes and classroom practices. Therefore, it might be that effective classroom pedagogic culture may vary with neighbourhood culture so that qualitatively different productive pedagogic cultures may thrive in different places.

Theories of identity can help us to understand how subjective engagement in classroom practices, through such opportunities for moments of identification, **constitute** learners' formation of identity, which may over time include a view of themselves as mathematical (e.g., as an enthusiastic learner of mathematics or a particular kind of maths person). For example, Gee (1999) elucidated the reciprocal relationship between language and identity. Identity is not only about being recognised or passing as a certain kind of person; it is also about (re)constructing oneself that way. In this view, when students engage or disengage with mathematics, they **also** thereby construct their 'selves' in practice. We may refer to this as identity in practice, and it is sometimes known as participatory identity (e.g., Boaler & Greeno, 2000; Wenger, 1998). In time, such participation may become realised and turn inward to become a narrative, self-identity (see Vygotsky, 1978; J. Williams et al., 2007).

Crucial then for this possibility is a pedagogy which openly encourages a peer-oriented mathematics talk. That is, if we are to impact or transform ways that students identify with learning mathematics, the development of a peer-aligned social mathematics holds the key—a way that, for some, may even enable a mathematical aspect of their identity to emerge. We note how Solomon (2007c) refers to mathematical identities as fragile. The role of pedagogy becomes to foster particular kinds of identification with mathematics within this social perspective.

CONCLUSION

In the extracts of classroom talk presented in this chapter, we did not recognise two separated peer and mathematics discourses, the one being of 'official maths', the language of the 'estimated mean', and the other (sometimes more quietly spoken and 'hidden') of everyday teenage talk. Such dual discourses can be described as a product of transmission-style teaching, whereby the division of labour places the teacher in a role as expert whose discourse (and 'classroom' tenor) is dominant at least when mathematics is voiced.

Our observational microdata show an alternative discourse, where the mathematics is social talk and a 'peer' social activity within the context of the mathematics classroom. We use Craig's 'crazy 20' to show how the peer sociality is appropriated into the co-production of a sociable mathematics. We argue that this social peer dialogue provides opportunities for points of connection and identification, and hence the potential for pedagogy to impact how students identify with mathematics. Such discourse practices are pregnant with potential for new worldviews.

We suggest that such acceptance of mathematics into the peer discourse practices of the students (even though this may remain situated within the context of the mathematics classroom) may be the first sign in a chain of acceptance of a view of themselves as mathematical. In other words, such discourse shows that Craig and Ellen were doing mathematics as part of their identity work, and we take this as a sign that, perhaps, they are accepting 'being a maths-person'.

Therefore, we argue that developing our understanding of such a social pedagogy for mathematics may be especially crucial with regard to widening participation in mathematics. It presents learners with the possibility of multimodes of belonging as the mathematics can align with a diversity of social discourses. We have argued that this is important regarding impacting certain ways of doing and talking about mathematics because a social pedagogy provides opportunities for change. We suggest, following Hicks, Moll, Gonzalez, and others earlier, that failure to provide opportunities for connection with students' lived and very meaningful social identities may be especially problematic for those who experience some dissonance within the dominant educational system.

We concur with Wenger (1998), who suggests that one of the most significant challenges faced by learners who move from one community of practice (or space for learning) to another is the reconciliation of 'forms of accountability' from those communities into one nexus. The extent to which students like Craig and Ellen can move comfortably between different communities of practice both in and outside college will depend on their abilities to coordinate the multiple perspectives to which they have been exposed. We note that the development of more inclusive practices will require attention at all levels of the education systems as more diverse practices may give rise to particular tensions should they need to perform within a more formal discourse of mathematics in the future.

ACKNOWLEDGMENTS

In this chapter, we have drawn on the ESRC TLRP project 'Keeping Opening the Door to Mathematically Demanding Programmes in FHE' (RES 139-25-0241), which investigates how pedagogic cultures impact on student's learning in mathematically demanding courses in further and higher education in Britain. This project examines how such courses can be enhanced to widen participation for those at risk of marginalisation and drop out from participating and continuing with mathematics (see http://www.tlrp.org/proj/wphe/wp_williams.html). The project design uses mixed methods, with the case study classrooms located within a larger longitudinal quasi-experimental analysis. We would like to acknowledge our project team and the contribution of the teachers and students who made this chapter possible.

13 Pedagogy, Discourse, and Identity

Stephen Lerman

INTRODUCTION

Under the influence of sociolinguists (e.g., Halliday), philosophers (e.g., Wittgenstein), psychoanalysts (e.g., Lacan), poststructuralists (e.g., Foucault), cultural psychologists (e.g., Vygotsky), cultural studies (e.g., Barthes), and others, the last (at least) 60 years of intellectual thought have witnessed a linguistic turn. By this is meant a view of cognition, affect, culture, knowledge, and its acquisition as being mediated by language. Sfard (2007a) uses the term 'commognition' as a device to distract readers from slipping into fragmenting language from knowing as two distinct but connected processes or aspects of consciousness and, instead, to see cognition, indeed consciousness as a whole, as identical with language and communication. Language precedes us, and we learn what things signify from parents, peers, teachers, and texts. Because language is culturally and temporally relative, as well as multiple across the communities within which an individual develops (gender, class, ethnicity, religion, race, physical location, sexual orientation, etc.), any individual is a unique collection of subjectivities. One can call this a strong linguistic turn, in that consciousness is seen to be constituted in and framed by language. Seeing language as affecting thought or as a tool to express thoughts might be called a weak role for language. One could refer to this shift also as a discursive turn. I take discourse to be more than language, whilst paradoxically it can only be carried in language (where else?). Discourse carries with it notions of regulation, of the power/knowledge duality of Foucault, and it is important to retain the connection, rather than view language as rather benign and neutral, a conveyor of thought that is somehow prior to and more essential than language.

The subjection of an individual to a discursive form is established through a pedagogic relation, which necessarily involves regulation through systems of power and control. Bernstein (2004) illustrates the nature of the pedagogic relation by giving two alternative responses a teacher might give to a child's drawing of a person with some body parts missing. He characterises an explicit or visible pedagogic relation by the teacher praising but pointing out that humans have two arms, two legs, and so on. He characterises an implicit or invisible relation by the teacher merely praising. These different relations will establish different forms of power and control. In my recent research, I have been working with the theories of Bernstein (2000) and have found they offer a strong framework (through a relatively strong grammar) (Bernstein, 1999b) for studying the regulation of learners (acquirers) in pedagogic relations (see Lerman & Tsatsaroni, 1998; Lerman & Zevenbergen, 2004; Morgan, Tsatsaroni, & Lerman, 2002; Tsatsaroni, Lerman, & Xu, 2003). Bernstein acknowledged the linguistic turn through his early study of Vygotsky, as one sees in these two quotes, one of them from the chapter on sociolinguistics in his last book:

> From Vygotsky and Luria I absorbed the notion of speech as an orienting and regulative system. (Bernstein, 1971, pp. 122–123)

> I came to the study of language by a diverse set of routes driven by the inadequacy of sociology to provide an orientation [. . .] I drew on the work in U.S. cultural anthropology, Russian work on speech as an orienting and regulative system (Luria and Vygotsky). (Bernstein, 2000, p. 145)

The mutual support provided by Bernstein's sociology and Vygotsky's cultural psychology is no accident. They were both inspired by Marx and, in particular, his perspective on the centrality of social relations in the formation of identity:

> It is not the consciousness of men that determines their being but, on the contrary, their social being that determines their consciousness. (Marx, 1859/1970, pp. 328–329)

Whereas Bernstein focussed on the access of people to symbolic control—that is, power by virtue of high-status positions in society (e.g., doctor, lawyer, academic) differentiated by the acquisition of an elaborated or restricted linguistic code—Vygotsky drew on Marx's statement to provide the mechanism through which culture determines consciousness. Vygotsky called for a unit of analysis that incorporated both cognition and affect, providing the possibility of accounting for human learning and behaviour that goes beyond Bernstein's strictly structuralist framework.

The book's theme is centrally concerned with explicating the notion of identity in relation to the teaching and learning of mathematics. Once again, it can be argued that there is a confluence of orientations toward the study of identity. First, it has become a common focus of attention in the social sciences in general. In 1996, Hall said, 'There has been a veritable explosion in recent years around the concept of 'identity' ' (p. 1), to which Bauman (2001) added, 'The explosion has triggered an avalanche'. I suspect that they are pointing both to the usefulness of the notion in studying human behaviour and to its somewhat excessive appearance in current writing, both academic and popular, becoming a fad or fashion. In popular writing, it has become more common to ask who one is or who one wants to be, rather than what one does or what one wants to do. The strap line of my university is 'Become who you want to be', a reference to identity as goals and aims, not what one wants to do or achieve.

One must look also at the sociological literature for the focus on life possibilities too in this period of late capitalism. In the literature of reflexive modernity (e.g., Beck, Giddens, & Lash, 1994), some writers claim that individuals are freer to write their own life scripts in this period than in previous periods that they call traditional and early or first modernity. In these earlier periods, people's identities were typically determined by family life and location, and by occupation and social class, respectively. Giddens and Beck particularly claim that now individuals can choose who they wish to be and are able to write their own life scripts, engaging in the project of the self, which is also termed by some as detraditionalisation. Beck tempers the freedom of choice that Giddens emphasises by talking of the risk society, in which one is forced to make decisions and changes by the inherent risks of the fast pace of development of society in late capitalism. In Boaler and Greeno's (2000) study, female students' choices not to go on to study mathematics at university—because the identity of mathematicians is perceived by them to be one that does not fit with their perception of their own identity or the person they wish to be seen as—is perhaps an example of identity work in late modernity. Others argue that structures, such as gender, continue to play dominant roles in identity formation and question whether we are in fact seeing a detraditionalisation. Instead, they argue that we are seeing a retraditionalisation, whereby old roles may be changed and extended, but do not disappear, particularly gender roles.

Second, Jean Lave, whose work has become almost ubiquitous in research in the mathematics education community (although used in ways that can be more or less productive; see Kanes & Lerman, 2008), argues for a shift in the language for describing learning from 'cognition' to that of identity.

> We have argued that, from the perspective we have developed here, learning and a sense of identity are inseparable: They are the same phenomenon. (Lave & Wenger, 1991, p. 115)

Looking at young people outside of schooling for the moment, one cannot but be aware of the manifestations of identity throughout their lives, whether it be in the clothing they wear in the desire to conform, to identify with a sports team or sports star or with a media star, or as expressed in the music they listen to, through which, in their choices, they express conformity to one group or another, or perhaps resistance to conformity. One notes that there are racial and cultural styles of dress, speech, and gestures which students may adopt, sometimes independent of whether they 'belong' to that social group. In the outward expression of religion, through dress, we again are strongly aware of identity and identification and the struggles that are engaged in by that form of expression. In the most recent times, young people are able to develop alternative lives in virtual worlds (actually people of any age, but this author is revealing his reluctance to engage with virtual worlds to this extent), where they take on identities and build environments and interact with others as and whom they choose.

In the school context, research studies of gender, ethnicity, and social class demonstrate the struggle for identity, acceptance, and, sometimes, just a peaceful path through childhood and adolescence, and through schooling in particular (see Kehily, 2001; Reay, 2002).

Looking briefly to research in mathematics education on identity, perhaps the most extensive work has been carried out by Boaler in a series of studies and publications (1997, 2002a, 2002b; Boaler & Greeno, 2000). Others include Bibby (2001), Mendick (2003a), Graven (2002), Bartholomew (2005), Sfard and Prusak (2005a) and Lerman (2006).

In examining identity, we must also take note of the effects that the regulatory systems of the state play in the identity of teachers and pupils. S. J. Ball (2001), for example, uses Baudrillard's notion of 'performativity' to describe the ways in which people are finding themselves responding to the dominating official regulation in the UK. He describes a self-regulation that differs from the panopticon of poststructuralism:

> Instead it is the uncertainty and instability of being judged in different ways, by different means, through different agents; the 'bringing-off" of performances—the flow of changing demands, expectations and indicators that make us continually accountable and constantly recorded. (S. J. Ball, 2001, pp. 211–212)

He quotes from Jeffrey and Woods' (1998) interviews to illustrate the impact of regulation on teachers' identities:

> I don't have the job satisfaction now I had once working with young kids because I feel every time I do something intuitive I just feel guilty about it. 'Is this right; am I doing it the right way; does this cover what I am supposed to be covering'. (p. 213)

> My first reaction was 'I'm not going to play the game', but I am and they know I am. I don't respect myself for it; my own self respect goes down. Why aren't I making a stand? (p. 215)

Pupils' identities in relation to mathematics are, of course, largely formed in classrooms, although at any time other aspects of their identities may come to the fore, as referred to earlier, and there may well be activities and interaction outside the classroom that play a part in their mathematical identities (Winbourne & Watson, 1998b).

I will now turn to an examination of the learning of school mathematics from the perspective of the formation of identities. I am looking here to discuss the identities we might want students to acquire in the mathematics classroom.

SCHOOL MATHEMATICAL IDENTITIES

I have used the term 'school mathematics identities' to acknowledge the recontextualisation that takes place when government, curriculum designers, textbook writers, and teachers construct the activities of the mathematics classroom. As Bernstein (2000) indicates, this process is always imbued with values—what one chooses as necessary, desirable, or appropriate for pupils to acquire. In this section, I offer some examples of classroom interactions that, it seems to me, illustrate the value of using the notion of identity to characterise the positions that students take up in the course of mathematical activities.

Example 1

Bernstein (2000) points out that students must acquire the appropriate recognition and realisation rules, knowing what things are when one encounters them and knowing what one is required to produce, in the pedagogic activity. He shows further that in student-centred classrooms, what he terms a 'liberal-progressive pedagogy', these rules may be hidden, to the disadvantage of students from working-class backgrounds. Children from middle-class backgrounds have generally acquired these rules in their home lives and know how to 'read' those tasks. For example (Lerman & Zevenbergen, 2004), the following task was taken from a series of interviews with students from a range of schools and year levels (Zevenbergen, 1991). The students were interviewed individually and asked to respond to the tasks:

Suppose you had a garden this shape and you were in a helicopter right above your garden looking down on it.

Which of the following shapes would be like yours?

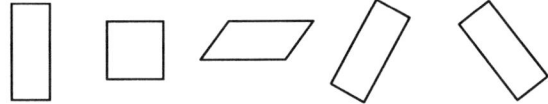

Although many of the students were able to respond to the task correctly, it was more likely that, when incorrect responses were offered, they were from students from working-class backgrounds. Typically, incorrect responses involved answering the question as if it were a task involving identification of the shape of their gardens at home. For example:

> *Robyn:* Why did you take that shape [the square]?
> *Girl:* Because it looks like the shape of my garden.
> *Robyn:* Is your garden at home like that?
> *Girl:* Yes.
> *Boy:* None of those.
> *Robyn:* Why aren't any of them the same?
> *Boy:* My garden goes like that [draws a semi-circle in the air].

Pupils are to recognise the task as one about shapes in mathematics, not about real gardens, and to know that the answer (i.e., giving the appropriate realisation of the task) is to come from within the language of mathematics. As Cooper and Dunne's (2000) study of students' performance on standardised test questions set in everyday contexts shows, social class effects are interfering with students' presentation of their knowledge by calling up the everyday rather than the required 'esoteric' mathematics. These students have not learned to read the task as a school mathematics task. Where students have not acquired the recognition and realisation rules, their identities do not converge to that of privileged school mathematics identities, but remain in the everyday. The elaborated language or code that middle-class students acquire in the home means that they can develop the privileged school mathematics identity more easily, seeing contexts as, for the most part, jokes or irrelevant distractions. Hasan's (2004) work on visible semiotic mediation or the informative mode in the home of middle-class children also supports this. Acts of localised pedagogy are present where the explicit transmission of 'knowledge' is made apparent in the discourse.

Drawing on the different discourses from home life and the forms of pedagogy as framed by the (mathematics) classroom, Bernstein's account provides an explanation of key aspects of the development of identities of teachers and students. It stops short, however, of affect and emotion (Adler & Davis, 2006), and hence does not provide a way of analysing other discourses and identities that may come into play for any student or teacher.

Example 2

The second example is not focussed on the disadvantage that working-class students experience in invisible pedagogies, but instead looks at what may be the effect of weak boundaries between mathematics and other school subjects, in this account design and technology in particular. As part of a research project to study 11- to 12-year-old pupils' mathematical problem solving[1], the following task was set:

Dear Design Team,

We are about to launch a new product onto the market. The product is a new breakfast cereal and we are looking for a design for the box which will contain the cereal.

There are several important criteria for the box: it must have a volume of 120cm^3 to contain the amount of cereal we want to include in one unit; it must be easy to stack on shelves in a supermarket and easy for customers to handle.

We would like the design of the box to be an appealing one. We would like the box to be as cheap to produce as possible, so we are looking for a design shape and one that prospective customers will find attractive which does not use too much card. If you would like to win the contract for this work please supply us with a brief report which should include your design for the box (including a scale drawing of the net of the box) and the reasons for the choices your team has made. We are looking forward to receiving your report and hope that we can work with you in the future.

Our (the teacher and the researchers') expectation was that the pupils would (a) try and construct cuboids or other familiar solids with the required volume, (b) soon realise that 120 cm^3 was far too small to be practical and would raise the problem with the teacher, (c) investigate the factors of 120, and (d) so on to other recognisably mathematical activities. In fact, almost all the pupils designed wonderful, colourful, unusual, and even bizarre shapes to satisfy the request for an 'appealing shape and one that prospective customers will find attractive', shapes whose volumes they (and we) could not easily calculate. Only one group of pupils realised the problem of the small volume, and they spent the rest of the lesson writing a letter to the company pointing out their error and asking for a new target volume.

In this case, although the weakening of the boundaries between subjects in school may have some advantages, and the potential for cross-curricular learning is desirable, pupils (and teachers) may have problems identifying the specifically mathematical. Bernstein discusses the problems for teachers' identities in what he calls an integrated curriculum, but this example indicates pupils' confusions too. Although what constitutes school mathematical activity, as distinct from other subjects in the school curriculum, is clearly demarcated, students (and teacher) can develop a clearer sense of what they are supposed to do in any activity and hence can be acculturated into a school mathematical identity in that space, at that time, as they are engaging in those activities. Where the boundaries are weak, the curriculum being built around common themes or projects, identities can become confused and blurred.

Example 3

Teachers will try to establish norms of interaction in the classroom that they deem appropriate. In the third example (Finlow-Bates, 1997), these first-year undergraduate computing studies students whose mathematics is not strong are engaging in problems on limits, a topic in their foundation mathematics course. Whilst the students travel down a false path, as teachers we may well be happy with the nature of their style of argumentation as being typical of mathematical thinking, involving conjectures, deductions, and refutations whilst at the same time showing respect for the opinions and ideas of others. They are working on the two infinite sums: $1/2 + 1/4 + 1/8 + 1/16 + \ldots$ and $1/3 + 1/9 + 1/27 + 1/81 + \ldots$

The next sentence in the conversation draws the group back together when Phi[2] [...] puts forward a bold 'naïve conjecture' regarding the sum of any series:

> *Phi:* Well. I think you double the original figure. Like a half will always be equal to 1. And your 1/3 will be equal to 2/3. That's what that means, isn't it?
> *Kappa:* No, it looks like it's converging to 1/2.
> *Mu:* But how come 1/2?
> *Phi:* It can't converge to 1/2.
> *Delta:* It can con ...[3]
> *Phi:* You're on a half there, aren't you? You're coming up to 1/243, 1 over ...
> *Kappa:* Yeah, I mean the numbers are so small, the numbers are getting smaller and smaller.
> *Phi:* It's the second term ...
> *Kappa:* And smaller.
> *Phi:* ..., that there makes sense to say that the sequence goes, converges to 2/3, doesn't it?
> *Kappa:* Yep?
> *Delta:* Yeah.

These students have acquired from their university tutor or from previous mathematics teachers a way of working with mathematical tasks in groups that conforms to the kinds of interactions that would be seen at least by the university tutor, who is also the researcher, to be desirable ones of mathematical argumentation. The tutor has an elaborated sense of what constitutes being a mathematician and is keen to induct students into the behaviours and ways of thinking required. Although the research did not engage with issues of how the students understood the nature of their activity, I suspect that they might well have described it in terms of what one does when doing mathematics. They could be said to have adopted identities of mathematicians, at least in the terms set by the tutor in the particular institution (Chevallard, 1992).

CONCLUDING REMARKS

What is gained by working with a notion of identity? Research on mathematics learning and teaching has always looked at the processes whereby students acquire knowledge, skills, and understanding. From the previous account of the role of forms of pedagogy and the determining function of discourse in the development of consciousness, what does 'identity' enable? In the prior examples, I have tried to indicate that students and teachers have aims, goals, and desires that they take into the mathematics classroom. For the teacher, and one hopes for some of the students at least, the main aim is to acquire school mathematics knowledge, sufficient for the goals of the students (to gain at least the minimum accreditation, to know enough for further study, etc.). That knowledge includes concepts and structures, conventions, processes, skills, pleasure, satisfaction and frustration, aesthetics, a sense of wonder, and perhaps other aspects of mathematical activity that have been recontextualised into school mathematics. I want to argue that 'identity' potentially incorporates all these into a unit of analysis, whereas other terms, such as 'cognition', 'achievement', and 'performance' are partial. Envisaging the process of induction of each student into the mathematics classroom as gaining a school mathematics classroom identity may focus the teacher's attention on the whole person and their becoming, rather than part of the person and their knowing. The attention of the teacher may thereby be drawn to the discourse of the mathematics classroom, with the recognition and realisation rules and the effects of forms of pedagogy on different groups of students. Finally, the teacher may become aware that identity is multiple. Although I have focussed here on the mathematical identity that will be the aim, students will have other overlapping goals and desires at the same time.

NOTES

1. British Academy Award No. SG 40073, 'Mathematical problem solving in Key Stage 3: researching mathematics in the classroom' co-directed by the author and Leone Burton.
2. The researcher is analysing the data using a framework drawn from Lakatos' (1978) 'Proofs and Refutations' and hence labels the students as he did using Greek letters.
3. In this text, '...' at the end of a sentence indicates that the next speaker interrupted and at the beginning of a sentence indicates simultaneous speech with the previous speaker.

Part V
Teacher Development

If you have been reading this book linearly, from cover to cover, you are now toward the end. After this part on teacher development, only the reflections in Part VI of Patricia George (Chapter 17) and Paola Valero (Chapter 18) remain. So, this last editorial introduction seems an appropriate place to comment on any emerging patterns within the three perspectives: sociocultural, discursive, and psychoanalytic. Then, after spending so long looking at the place of identities in discussions of mathematics education, we turn to wondering about the place of mathematics in discussions of educational identities.

This part opens with a psychoanalytic response from Pat Drake (Chapter 14) to the following question: How do initial teacher training and professional development programmes impact teaching and learning practices within educational contexts and, subsequently, learner identities in relation to mathematics? She draws on Lacanian ideas to make sense of her student teachers' relationships with mathematics. Following Lacan, Pat makes different assumptions about the self from those operating within the existentialist approach of Mark Boylan and Hilary Povey or the object relations approaches (drawing on Klein, Bion, and Winnicott) taken by Jenny Shaw, Tamara Bibby, and Melissa Rodd and ourselves in our chapters. However, we highlight here two features we feel distinguish these chapters as a group. The first is a close attention to the emotional and a commitment to the idea that much is happening 'beneath the surface'. Pat, in arguing that we need to take account of new teachers' emotional as well as intellectual relationships with mathematical knowledge, offers a significant challenge to how we conceptualise subject content knowledge, pedagogical content knowledge, mathematics for teaching, or whatever else we label the things we (think we) want mathematics teachers to know. The second commonality is a holistic, biographical, and narrative approach. Psychoanalysis seems to draw us to stories—ones that stay with the data, but that disrupt familiar, commonsense readings. Further, there is a pull to the autobiographical: In Pat's chapter, as in Jenny's and Laura et al.'s, we learn much about the writers' own relationships with mathematics.

Following Pat's chapter, Barbara Jaworski (Chapter 15) offers an account of a teacher development programme based in Norway, in which didacticians and teachers worked collaboratively. It is possible to see connections with the sociocultural analyses of Rachel Marks and Jeremy Hodgen, Pauline Davis and Julian Williams, Birgit Pepin, and Peter Winbourne. In these chapters, there is a focus on communities and/or figured worlds, on the practices that take place there, and on whether and how students and teachers participate in these. (Not) belonging becomes a way of understanding what possibilities there are for learning and teaching within a particular space. Barbara draws parallels between what goes on in mathematics classrooms and what goes on in professional development work. In this way, she raises an often ignored question—What is teaching?—and she considers the relationship between teaching and learning. She gives an open account of the tensions involved in developing a community of didacticians and teachers from multiple institutions, stages, and phases. She highlights both the ways that their roles overlapped and the effects of power imbalances between them. She describes these power imbalances reducing or even disappearing as understanding grows between participants, through processes of imagination and alignment.

Finally, Tansy Hardy (Chapter 16) analyses regimes of truth in which mathematics teachers come to understand themselves. Tansy's chapter, like the other discursive contributions to this book—by Anna Sfard, Steve Lerman, Candia Morgan, and Heather Mendick, Marie-Pierre Moreau, and Debbie Epstein—scream power. In contrast with Barbara's construction of power as problematic—something to be eliminated—here it is unavoidable and ever-present. Although power jumps out from these chapters, the role of identity is, perhaps, less obvious. However, it is there in the way that the self is formed in discourse—the knowledges through which we come to know ourselves and be known by others. Tansy's students experience themselves to be confident or not, competent or not, and qualified or not through the available discourses. She shows how policy and research can never in any simple sense identify, analyse, and solve a problem because these very acts constitute the problem they pretend to describe. As she puts it: 'The field of teacher development in mathematics needs to entertain the possibility that it is the very conceptualising of primary teachers' professional knowledge of mathematics which, in its articulation, generates and condemns teachers to having faulty knowledge'.

This leads us to the place of mathematics within constructions of educational identities. It is commonly assumed in the literature on the professional development of mathematics teachers that 'the added complication in mathematics teacher education pedagogy is that there are two objects in play: mathematics and teaching' (Adler, 2008, p. 2). In England, this has led to the government establishing a high-profile, subject-specific organisation to coordinate teacher development in mathematics: the National Centre for Excellence in the Teaching of Mathematics (NCETM). The transparency of the need for such provision is apparent in the way that there is no justification offered for the NCETM's existence on its website (www.ncetm.org.uk).

But any focus on mathematics needs justification. We need to ask: What difference does mathematics make? In the context of this volume: How would a general book on educational identities differ from this one? These are complex questions which were with us often during the seminar series that preceded and precipitated this book, as well as through the process of editing. We do not attempt to answer them here, but instead to take advantage of the opportunity to juxtapose what Tansy, Barbara, and Pat have to say about these questions. It is juxtapositions such as these, and the ways that they can open up thinking, that are behind this book's structure.

Pat argues that there is something about mathematics that compels us to teach and learn the subject and which locates it, in Lacanian terms, 'in the place of 'the rule bound symbolic Other' ', institutional authority, the 'law of the father'. Judith Butler (2008) has written that: 'any child has access to a range of masculinities that are embodied and transmitted through a variety of cultural means' (p. 7). Pat locates mathematics as one such cultural means. Barbara points to the way that, in mathematics, 'many concepts are not available to children through everyday activity'. This preponderance of scientific concepts, she suggests, demands pedagogic interventions by the teacher that are different to those required in other subjects. Thus, both Pat and Barbara confer a special status onto mathematics as different from other subjects, as indeed does this book (although in Pat's case, this conferment is provisional, 'for the sake of argument'). Tansy draws attention to how these understandings are constitutive of regimes of truth which set mathematics up as different: as in need of such books, as a difficult and problematic subject, and so on. As Judith Butler (2008) puts it 'the text does not have an epistemically privileged relation to its subject. It is part of the project to compel the production of that subject, and we will have to ask why' (p. 16).

14 Mathematics for Teaching
What Makes Us Want to?

Pat Drake

INTRODUCTION

We all recognise the stereotype of the fusty, scruffy, highly eccentric, frequently male teacher of mathematics, the person who is deeply unfashionable, has dubious standards of personal hygiene, and has few friends. I am such a person, a female version, and I have spent much of my working life trying to avoid this unpleasant labelling. If being a mathematics teacher means living with a stereotype of social unacceptability, what on earth makes us want to do it?

I have come to believe that something compels us to teach and learn mathematics in ways that are crucially linked to our identities, to who and what we keep faith with, and that these matters are deeply individual. For as Buxton (1981) wrote: 'There is a nakedness about the success or failure in reaching a goal that invokes clearly defined emotions whose nature one cannot disguise to oneself' (p. 59).

In this chapter, by drawing on the voices of a range of people becoming teachers of mathematics, I explore the relationships between their emotional engagement with mathematics and a desire to teach it. I address these questions:

- Who becomes teachers of mathematics?
- Why do they/we engage with mathematics?

I draw on accounts from two specific groups of new mathematics teachers: people with mathematics or related degrees beginning a postgraduate initial teaching qualification in England (PGCE), and some people without any previous advanced mathematics qualifications at all who have decided to train to teach mathematics through their experience of supporting mathematics learning in secondary schools (ages 11–16 or 11–18) as teaching assistants. Through these accounts, we can see how people's relationships with mathematics connect with their identities in various emotional ways, and that variations in these relationships are very likely to lead to different approaches to teaching. This suggests that when working with new mathematics teachers, we should take account of their emotional relationship with mathematics knowledge, as well as their intellectual grasp of it. However, initiatives to develop mathematics teachers are frequently couched in terms of mathematics subject knowledge (see e.g., TDA, 2007).

The chapter begins with part of my own mathematical story. I then introduce a psychoanalytic theory from the work of Jacques Lacan to show how mathematics and meaning-making are connected emotionally, and to illustrate this theory using extracts from my data.

WHO ARE WE MATHEMATICS TEACHERS?

As a secondary school student, I was deemed good at mathematics, rather than being especially interested in it. Thus, it seemed an easier A-level option than the other choices that required more work, and I was able to skate through to mediocrity with ease, despite the entreaties of my dedicated mathematics teacher who tried her utmost to persuade me to adopt a 'deep learning' approach (Entwistle, 1981). I pursued this superficiality through university, where unfortunately even mediocrity eluded me and from where I emerged on my hands and knees with a very poor degree result indeed, having also lost all confidence in my mathematical capabilities.

I went into mathematics teaching. (Were I now to apply to be a mathematics teacher with the qualifications I had when I was 21, I might not be accepted.) Much to my surprise, I developed a love of teaching mathematics, including to A level, and I began to engage with the school mathematics that I had previously disrespected, but more warily and with little confidence. Some years later, I was able to redeem myself academically through the generosity of the late Laurie Buxton, at the time Chief Inspector for Mathematics in the Inner London Education Authority.[1] Laurie encouraged me to take a sabbatical year to undertake a full-time postgraduate qualification in statistics, and he trusted me to make the most of it. This time study seemed easy and enjoyable; four of us on the programme worked together to make sense of the material, and we all did very well in the exams.

I now believe that I understand most of the mathematics I know through teaching it or through working alongside other people.

Practicalities regarding the provision of mathematics teachers are in tension with what is seen as desirable in terms of high-status mathematical knowledge. In England, demographic trends and policy of expansion, the place of mathematics in the National Curriculum, and a shortage of well-qualified teachers mean that non-mathematics specialists teach many school students, drawn in with varying degrees of enthusiasm as the need arises. This is borne out by evidence from a sample of schools presented in the *Secondary Schools Curriculum and Staffing Survey* (Department for Education and Skills, 2003) and the reaction by the then Secretary of State for Education, Charles Clarke. According to the data, 26% of those teaching mathematics had no qualification in mathematics, and 14% of mathematics lessons were taught by teachers with no qualifications in mathematics. Charles Clarke (quoted in Guardian, 2003) said:

> A proportion of maths teachers are listed in the Survey as having 'No Qualification' in maths but this doesn't mean they are unqualified. Most of these teachers are likely to be qualified and graduates in subjects such as physics and ICT. They may only teach one or two periods of maths a week.

In fact secondary school mathematics has by no means the highest numbers of people who are teaching it with no qualification in the United Kingdom—it is 16th in a list of 28 subjects with, for example, Information and Communication Technology, Religious Education, Drama, Business Studies, Spanish, German, and Geography all ahead in terms of having fewer people teaching qualified in those subjects. So teachers of mathematics are not unusual in teaching without formal mathematics qualifications—many subjects are in the same position—and many teachers of mathematics teach on the basis of their professional knowledge and sympathetic understanding of learners, with graduate or postgraduate status, but frankly with very little conventional evidence of high mathematical achievement.

I have worked with several different groups of such people teaching mathematics (see e.g., Drake, 2001, 2005). As well as conventional mathematics graduates, these include teaching assistants and teachers retraining to teach secondary mathematics from other subjects, as well as those with professional allegiances elsewhere (e.g., nonspecialist higher education teachers who support undergraduates' mathematics and/or statistics needed in disciplines such as economics or geography) or those who find themselves on the sharp end of government policy, such as supervisors on various training schemes confronted with the need to teach mathematical skills explicitly in the context of workplace training.

Given the paradoxical place of mathematics in the curriculum, being seen simultaneously as an easy option for those with recognisable aptitude, yet seriously forbidding for others, over time I have become interested in why and how people with and without conventional mathematical 'qualifications' are drawn to try to support others' mathematical learning.

In my current institution, graduates starting their professional teacher training are asked to write and send ahead of their arrival an autobiographical account of their education to date. People respond to this request in different ways, and over time I have collected written accounts from trainee teachers of mathematics that contain a wealth of detail regarding school and university experience, schooling in the UK and overseas, friendships, expectations, family relationships, relations with teachers, and relations between teachers and subjects and between the individual and their perceptions of mathematics.

I also recently completed a study of secondary mathematics teaching assistants undertaking a 3-year honours degree in mathematics education (Drake, 2006). As part of this study, I conducted a series of interviews with the teaching assistants about their experiences of learning mathematics in the university and in the schools in which they worked.

WHY DO THEY/WE ENGAGE WITH MATHEMATICS?

Some Psychotherapeutic Notes as Background

A psychoanalytic theory that is beginning to be used in considering teacher preparation and development is from the French psychoanalyst Jacques Marie Emile Lacan (1901–1981). This is because Lacanian theory proposes a triangle of relations between our prelanguage selves, called our Real selves; our selves that we construct as a result of our relations with other people, which Lacan says is our Imaginary self; and our relationship with what he calls the 'Symbolic Other', that is to say, the rules and structures of the society in which we live. Teachers in England have to operate within the structures of a set of defining competence standards, arguably Symbolic, and Other (see e.g., Hanley, 2007; Hanley & Jones, 2007), and it is teachers who help mediate the subject, such as mathematics, for learners.

Lacan articulates a position for us whereby we are always in flux, driven to make meaning arising from the inevitable dissonances between ourselves as defined through other people such as family, ourselves as we are in relation to authority, society's rules and regulations, and ourselves as we in essence are. Lacanian theory proposes that dissonance arises because language structures who we are in relation to other people and forces a disjuncture between our 'Real' prelanguage selves and these others. Our identity develops as we 'mirror' ourselves through language with others. Lacan says this creates an 'Imagined self' and is an illusion that we can get what we want from other people. However, it is a misrecognition, and so each of us is troubled by a sense of loss. People are also structured through language with respect to structures of society, such as laws and institutional practices. Lacan calls this the **Symbolic** order. The child has to learn to desire beyond his or her mother, just as the mother desires beyond the child, and what is signified by this desire is in psychoanalytic terms, the father, or the phallus, a symbol of potency and power. The signifier, having now moved beyond the object to a representation of it, becomes a 'deferred signifier', and Lacan says 'the name of the father', a potent **Symbolic Other**, represents institutional authority, or 'the one who knows'. Thus, the person is connected to the structures of society, the laws and regulatory practices through their learning to desire beyond their mother and, in so doing, accepting a symbolic 'one who knows'.

As people we continually seek to make good our loss; driven by our desire, our compulsion to try to find that which cannot be satisfied, we are provoked into actions, be these personal or professional. If we understand this meaning-making process, then we understand ourselves. Žižek 2006) explains that 'the goal of Lacan is not the person's well being but to bring the person to confront the elementary coordinates and deadlocks of their desire' (p. 4). This is our Real self, but because it is not structured through language, it is not attainable. In the Lacanian sense, meaning is not fixed, but is exploratory, and it provides a creative sense for us as we develop a sense throughout our life of one meaning leading on to another.

According to Lacan, we are all naturally hysteric or obsessive. Hysterics are concerned primarily with the question, 'Who am I and what do I really want?', and obsessives want to escape being dependent on 'the one who knows'. If desire becomes blocked, the person becomes stuck, and the purpose of analysis is to restore desire or a searching for meaning. Lacan ascribes gender differences to hysterics and obsessives, with females tending toward hysteria and males more associated with being obsessive. Whichever way we are oriented, healthy people defer meaning and work toward an acceptance of loss in life, recognising that no one really knows, and that in terms of knowledge there is no deferred signifier. Healthy people work to create more interesting meanings from what they 'understand' already.

In this chapter I suggest, for the sake of argument, that we put mathematics, the subject, in the place of 'the rule-bound symbolic Other' because this is the way many people see it. This means that, from a Lacanian point of view, mathematics learning can be seen as a movement between meanings, with making mathematical meaning lying with the student and not with the teacher. This would suggest that 'know-it-all' teachers set up dependent relationships, whereby the teacher replaces the subject as the 'one who knows'. This is off-putting for many who create situations whereby, if they cannot be like the know-it-all, they do not want what he or she is providing. Does this help us think about mathematics teacher development in a helpful way?

In the next section, I use extracts from the accounts of three new graduate trainee teachers of mathematics: Andrew, Pete, and Janet (pseudonyms). Pete and Janet were mature entrants, and Andrew was a recent graduate. They talk about their relationship with learning mathematics in various ways, and I try to link what they say with a Lacanian perspective on making meaning about mathematics and, consequently, their identity work.

After this come extracts from the stories told by secondary teaching assistants (pseudonyms Barbara, Caroline, Carrie, and Margaret) supporting school mathematics whilst simultaneously undertaking a degree programme that allows access to mathematics teacher training.

166   *Pat Drake*

In learning mathematics, knowledge structures of the subject are not value-free, to be learned simply as 'tools of the trade'. It is likely that acquisition of the knowledge itself is a site of struggle for position, influence, status, and recognition. The second examples illustrate a group of individuals hitherto excluded from mathematics, but who set out to engage with it. This was not therefore a transformation of usual novice mathematics students, but of a group previously excluded from mathematics learning.

Traditional Mathematics Qualifications

Andrew, who obtained a first-class degree in physics, never doubts his ability: It is just a question of working harder, although his choice of mathematics is self-declaredly to avoid work. He talks about his choices of subjects to study for A level between ages 16 and 18:

> *Andrew:* I had decided that because of the nature of maths homework (lots of short solutions to problems as opposed to long essays) it would lend itself excellently to doing at the bus stop and on the bus to and from college. This would save me from doing any work in my own time. The natural subject to accompany double maths was of course physics as the syllabuses overlapped quite a bit resulting with less overall work to do. So, armed with my little-work-as-possible choice of subjects I started sixth form [ages 16–18] with the full expectation of a year or so of leisure and then a couple of months of cramming for the final exams.
>
> I remember drifting through most of my first year [. . .] then three things happened that totally changed my approach. The first was that I became a Christian [. . .] if I was doing it for God, then just getting by was not really an option. The second was that we were encouraged to start looking into going to university, and I was shocked to discover that the University of Birmingham would not be satisfied with the assortment of [grade] Ds and Es as I was currently getting. Third, I was finding it harder and harder to 'get by' because [. . .] as I didn't understand yesterday's work I found it almost impossible to grasp what we were doing today.

Andrew is in control. Analytically, he would be obsessive, dependent only on the most metaphysical 'one who knows', the very highest authority he can find, God. His only concession to material authority is the generalised 'University of Birmingham', who would 'not be satisfied with the assortment of Ds and Es as I was currently getting'.

Pete had a degree in Computer Science and about 5 years of experience in the industry. He explains his approach to studying, reflecting on his assessment of himself as a member of a class:

> *Pete:* I suspect that I wasn't a particularly interesting student to teach. I was probably a very poor contributor to lessons. My approach to learning was a solitary one. As a result I rarely remember getting much attention from the teachers, and now no teacher stands out in my recollection above any other.
>
> During my secondary school years my solitary but strongly self-motivated approach to my own learning became more of a feature. In school, after taking my maths O' level [examination that used to be taken at age 16] in November I spent 6 months studying for a statistics O' level from worksheets before doing well in the exam. And at home I spent hours unravelling the deepest secrets of my computer. Although I hope that since this time my experiences have helped develop a more open and group oriented approach to learning, the ability to quickly teach myself something from a book has stayed with me and proved invaluable. [. . .]
>
> The change from 2 year O level course to in particular a 1 year Maths A level was a significant step up. I was fortunate enough to be taught at this time by the most outstanding teacher of my academic experience, Mr Looward [. . .] a definite team atmosphere developed in class and for the first time I started to move away from my predominantly solitary approach to learning. [... At university] the course was well planned and run by a very professional department [who] gave the impression of being at the leading edge of the subject.

Pete studies statistics on his own, and 'spent hours unravelling the deepest secrets of my computer' defining himself at that time as 'solitary'. In Lacanian terms, Pete would also be seen as obsessive because he only allows himself to be dependent in situations where he acknowledges and trusts the excellence (beyond him) (e.g., the outstanding Mr. Looward and at university).

Janet came into teaching in her mid-30s. Her tale offers a complex piece of writing, as she weaves her family relations into her account of her school and university experience of mathematics:

> Apart from my mathematical education I have very few memories of other subjects or their teachers. I learned in the shadow of my sister who is my elder by 4 years; and felt that if she couldn't do something then it would be pointless for me to try. Fortunately I also followed

> my sister's strengths (primarily mathematics) but I was never able to escape the mental block of feeling I would only ever fall short of her accomplishments. [. . .]
>
> I did encounter one very effective source of pressure. His name was Martin Simmonds. When we were streamed and introduced to the boys in our class I found that I had really met my match. Mr Clutton would regularly set us tests in class, and when the results were announced one or two lessons later, my adrenalin levels would be sky high. I HAD TO BEAT him.
>
> About the time I was taking my GCEs [O level exams] my sister was nearing the end of her first year at Cambridge. I had just sat the final exam when the bombshell dropped. My sister had failed her first year exams. [. . .] My GCE results were very good, better than my sister had done. I wanted to move to a different school to study for A-levels and this was my opportunity.

Janet's account of competition and comparison with others apparently finds a legitimate place in her study of mathematics, although is likely to be driven from other desires and interactions. She describes her teacher:

> Our teacher [in the new school] came across as if she too were working out the answers for the first time and had as much to learn as us. We would end a lesson not having discovered an answer and come to the next lesson hoping to be the one to have solved the problem in the meantime. Mrs Matton recounted how she always slept with a notepad by her bed, as sometimes the answers would come to her in the middle of her dreams. I empathised with this as it often happened to me.

In Lacanian terms, Janet is hysteric because she locates 'the one who knows' (i.e., mathematics knowledge) as very powerful, but outside her immediate control, in the mystical domain of her dreams. This is unlike Andrew and Pete, who apply themselves manifestly in physical realms. She also refers to university in a different way from them:

> I was fulfilling no real goal by going to university except to occupy the 3 years that [my boyfriend] Jimmy would be away [at university]. When Jimmy and I visited each other our different study techniques became blatantly obvious even though we were studying the same subject. Jimmy worked a lot harder than me. He would spend a greater time trying to resolve a tutorial problem, even researching external material where necessary. If I couldn't complete the tutorial exercises using lecture notes I would study the answer sheets given out after the actual tutorial class. Jimmy's secondary schooling had obviously trained him

to work independently and he never felt it a hardship to go back to first principles to work something out. The depth of his understanding was far greater than mine.

Without her story of her sister and her father, might we interpret Janet's apparent competitive approach to mathematics and her reliance on performance goals in other ways? What are we to make of her delight with her sixth-form teacher, Mrs. Matton? How are we to explain her comparison with her boyfriend ('The depth of his understanding was far greater than mine')? Janet is seeking completion by an Imaginary other. Her relationship with her father is such that she will not get her needs met, so she replaces her father by her boyfriend as 'the one who knows'.

Nontraditional mathematics qualifications

Now we look at a small group of 10 secondary mathematics teaching assistants, 9 women and 1 man. Despite starting with relatively low qualifications in mathematics, eight of them went on to become qualified graduate teachers of mathematics for students ages 11 to 16. All had acquired additional qualifications since leaving school in leisure pursuits such as embroidery, in technical skills such as computer literacy, and/or in professional domains (e.g., teaching basic skills as well as more academic qualifications). Each was used to 'being a learner'.

What prompted these students to want to pursue mathematics so as to train as teachers? Caroline explains her relationship with mathematics differently as an adult than as a school student, with her desire to make sense of it stimulated in relation to her daughter's desires.

> I only really got interested in maths when my daughter, who is now eighteen, started here, and brought some homework, and asked for some help, and I couldn't help her. And then I noticed that the school ran an adult basic skills class. So I originally came to get help with my daughter's homework, and from that I got interested in maths, and I took an adult achievement award, and then I moved on to [two higher qualifications]. I just found that I, you know, quite like maths. It was interesting and a completely different experience than I had when I was learning it for myself.

Margaret makes sense of her intellectual ambition in relation to other trainee mathematics teachers.

> I'd always done an evening class or something, and always I'd been getting a bit of training through work and to be honest some of it was damn easy, well I wanted to do something a little bit more advanced.

> I got on to the Diploma and enjoyed it so I thought I'd give it a go. And there was an element of there not being many maths teachers, and seeing people coming in and thinking I'm sure I couldn't do any worse than they do. I might not be able to do any better but I don't think I could do any worse. It sort of spurs you on.

Many of the students had liked mathematics at school themselves, although several had been actively discouraged from pursuing it: 'I went to a grammar school, I wanted to do A-level maths and was told I wasn't good enough and I wasn't allowed to do it', said Barbara, thereby locating 'the one who knows' in her teachers. Connecting school mathematics experience with teachers in preference to the subject is evident also in what Carrie says: Despite her enduring enjoyment of mathematics, her expectations of her performance are always in relation to the teacher—that is, whether the teacher resonated with her Imaginary other.

> I used to quite like it at school, and I was quick to learn things, by rote, times tables, certain formulas, etc. etc, and I used and still do like playing around with numbers. The second year was this bloke, he was horrible, the third year was a really nice bloke again, it depended on the teacher as to how I got on. And then we had the git again for my O-level, and I think I could have done a lot better if I'd had one of the other teachers.

Whilst at school, these students had embodied 'the one who knows' in their teachers. However, as adults, they had shifted so that their meaning-making came out of their experiences and desires in relation to both other people—'I don't think I could do any worse'—and the mathematics embedded in the practices at the schools where they worked.

DISCUSSION AND CONCLUSION

The extracts, chosen to contrast with each other, show the authors' subjectivities being expressed in relation to their mathematics experience along different continua: in control/out of control, dependent/independent, material/metaphysical/mystical, social/solitary, real/dreamlike, emotionally neutral/emotionally loaded, connected with others/disconnected from others, competitive with others/competitive with self, and not motivated/intrinsically motivated/extrinsically motivated.

Drawing on such a well-developed field as psychoanalysis in order to illuminate another area entirely is tricky because the complexities of meaning-making through psychoanalytic work are the preserve of the analysed and the analyst, and I am neither of these in any of these cases. However, some of Lacan's ideas provide a means of looking for ways to think about what it might mean emotionally for beginning mathematics teachers to be engaging with mathematics and how this just might play a part in shaping their identity as teachers.

The first group of students are different from the second insofar as they had not yet started to work in schools, whereas the students in the second group were working in schools that motivated their desire to make meaning out of mathematics and out of teaching mathematics. There is irony here because the very classrooms that stimulated this desire are the location for alienation of many school students for whom the school curriculum is designed. For the first group, mathematics in school provided an opportunity to be successful, and it is to the classrooms that alienated their peers that the students return.

For the second group, transformation occurred through their extending a sense of self with respect to mathematics practice in schools rather than through a radical redefinition with respect to mathematics itself. What comes through from these students, in a similar way for Andrew and Janet but not so much for Pete, is the emotional investment that the students placed on being successful and the emotional work that was incurred in being successful.

The second group of students exemplify a different type of mathematics learner—those for whom mathematics has never been easy, who have never been recognised as talented, and who have developed as a consequence successful strategies for dealing with the almost inevitable difficulties. For them, professional progression combined with personal disposition and serendipity is what brings them to teaching mathematics. In so doing, their arrival forces us to think about what is important about mathematics knowledge acquisition.

I'm not trying to point to generalities or similarities between the individuals or between these groups. However, in the current climate of mathematics teacher shortage, whatever the evident level of previous mathematical attainment, I think those of us involved in the preparation and development of mathematics teachers might find that, in addition to developing mathematics subject knowledge, thinking about the following from an emotional perspective is also helpful:

- individuals' perception of their own performance in mathematics, and how this relates to their experience of learning
- relationships between teaching and learning mathematics
- enjoyment/pain of (un)successfully dealing with difficulties
- significant other people in learning mathematics
- competition and comparison
- self-motivation and extrinsic motivation
- auto-didacticism, self-control, and independence
- social class, gender, and opportunity

ACKNOWLEDGMENTS

I would like to thank the organisers of the ESRC Seminar Series Mathematical Participation and Identities in 2006–2007, Jan Campbell, all of the mathematics teachers who have been quoted anonymously, and my teachers Miss Briggs and Miss Edwards because both of them taught me some mathematics at school that has lasted for life.

NOTES

1. The Inner London Education Authority was disbanded in 1990 when local responsibility for education in inner London was divided among a number of smaller boroughs.

15 Developing Mathematics Teaching Through Collaborative Inquiry

Barbara Jaworski

INTRODUCTION

This chapter addresses theory and practice in a developmental research project in which mathematics teachers and didacticians worked together to develop mathematics teaching. The mathematics teachers were from eight schools in Norway, ranging from lower primary to upper secondary. Didacticians were academics in mathematics education in a university. Both were practitioners in their own areas of practice, and in the project both were researchers. The project sought to know more about how mathematics teaching can develop to enhance learning experiences for students in mathematics classrooms. It was called Learning Communities in Mathematics (LCM)[1] and involved teachers and didacticians in inquiry communities exploring together and evaluating possibilities for classrooms and students (Jaworski et al., 2007). Over a period of 4 years, including three phases of school-related activity each of one school year, 14 didacticians (including five doctoral students) and 35 teachers were involved in the project.

The theoretical root of the project was the notion of **inquiry community**—of mathematics teachers and didacticians exploring together and evaluating possibilities for classrooms and students. The project was conceptualised around **inquiry** in three levels or layers:

1. in classroom mathematics,
2. in planning/designing for the mathematics classroom, and
3. in researching the developmental process.

Inquiry was seen as both a tool in promoting mathematical thinking and central to both development and research. We see the three elements as deeply related to each other in a sense of nested layers as represented in Figure 15.1. Research (outer layer) into the activity of the two inner layers both charts that activity and its development while simultaneously contributing to that development (Jaworski, 2003).

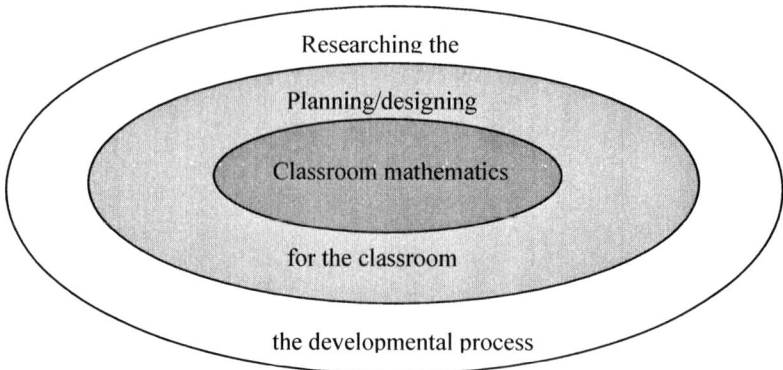

Figure 15.1 Nested layers of inquiry in the LCM project.

LEARNING THROUGH PARTICIPATION AND THE POSITION OF TEACHING

Community of inquiry draws on Lave and Wenger's (1991) theory of **community of practice** and Wenger's (1998) construct of **belonging** to a community of practice. **Community** refers to a group of people identifiable by who they are in terms of how they relate to each other, their common activities and ways of thinking, and their beliefs and values. According to Wenger (1998):

> The concept of practice connotes doing, but not just doing in and of itself. It is doing in a historical and social context that gives structure and meaning to what we do. In this sense practice is always social practice. (p. 47)

Wenger (1998) suggests that **belonging** to, participating in, or having identity in a community of practice involves **engagement, imagination,** and **alignment**. We **engage** with ideas through communicative practice, develop those ideas through exercising **imagination**, and **align** ourselves 'with respect to a broad and rich picture of the world' (p. 218). **Align**, literally 'to line up with', indicates that we are positioned according to, or in line with, the practices and activities in the communities in which we participate.

The terms **participation, belonging, engagement,** and **alignment** all point to the situatedness of doing and being and the growth of knowledge in practice. In LCM, for example, teachers and didacticians **engage** in practices in workshops and school settings and **align** themselves with existing or emerging practices related to the particular setting. **Imagination** contributes to the emergence of new practices. Interactivity and relationships among people within a community establish norms of practice and **ways**

Developing Mathematics Teaching Through Collaborative Inquiry 175

of being in the community. Identities are formed through situated engagement; individual or group identity relates to ways of being in communities to which we belong. What people do, what they know, and their concomitant growth of knowledge relate to their interactivity in social settings. For example, in LCM, which had teacher teams in eight schools, any individual teacher- or didactician- developed identity related to their participation in the project particularly, but constituted relative to the many other communities of which the individual was part.

Thus, within a theoretical perspective of community of practice, participants are seen as situated with respect to their practice and learning through participation in practice (Rogoff, Matusov, & White, 1996). Because participants are teachers (and, to an extent, didacticians are also teachers), we need to consider how the roles of teachers can be seen to fit within such a community learning model. Therefore we ask, **if learning is participating in a community of practice, what is teaching?** If a teacher is supposed to enable **others** to learn, how can a teacher enable learning through participation? What can this mean?

Lave (1996) writes:

> People who have attended school for many years may well assume that teaching is necessary if learning is to occur. Here I take the view that teaching is neither necessary nor sufficient to produce learning, and that the socio-cultural categories that divide teachers from learners in schools mystify the crucial ways in which learning is fundamental to participation and all participants in social practice. (p. 157)

Social practice theory is illuminative in offering a means of characterising and analysing learning (e.g., teachers' learning of mathematics teaching or pupils' learning of mathematics). However, the frame is unhelpful in characterising or analysing mathematics teaching—that is, enabling **others** to learn. Indeed, according to Lave, 'teaching is neither necessary nor sufficient to produce learning'.

Children learn through participation in many contexts, both inside and outside the classroom. Where mathematics is concerned, however, many concepts are not available to children through everyday activity. Mathematical concepts are scientific concepts (Vygotsky, 1962) and, according to Schmittau (2003), 'require pedagogical mediation for their appropriation' (p. 226). Pedagogical mediation involves the (a) creation of a classroom environment in which learners can engage with mathematics and encounter norms of mathematical practice, and (b) provision of scaffolding related to zones of proximal development (Bruner, 1985). Such provision seems to sit firmly within the role of a teacher of mathematics.

So, how do we interpret the term 'teaching'? What exactly is taken to be the role of a teacher? To see mathematics learning as through participation in a community of mathematical practice, we have to consider how such a community is created. For example, we might see mathematical practice, 'doing mathematics', to involve elements of mathematical tasks, mathematical thinking, mathematical reasoning, generalisation, abstraction, and proof. Where can such elements be found or how can they be created in classroom situations so that pupils can engage with them? These questions provide a possible way ahead: If we see learning mathematics in classrooms as participation in the social practice of the classroom, we might see teaching mathematics similarly as participation in the social practice of creating opportunity for mathematics learning. In both cases, inquiry can be seen as an important tool in the process: Inquiry promotes questions into mathematics or into mathematics teaching which, concomitantly, encourage deep levels of participation and reflection.

Lave (1996) writes further:

> if teachers teach in order to effect learning, the only way to discover whether they are having effects and if so what those are, is to explore whether, and if so how, there are changes in the participation of learners learning in their various communities of practice. If we intend to be thorough, and we presume teaching has some impact on learners, then such research would include the effects of teaching on teachers as learners as well. (p. 158)

If we see teaching as a practice in established settings, aimed at promoting learning of mathematics, then we might simultaneously see learning of teaching through participation in teaching. This suggests a developmental model for teaching. However, research of Brown and McIntyre (1993) suggests that such a model is not obvious as a part of normal teaching practice. In their study of teaching in natural settings of classrooms over a substantial time period, Brown and McIntyre saw the teaching-learning situation settling down to norms that they referred to as 'normal desirable states'. A state was **desirable** in that teacher and pupils found ways of working together that were comfortable to both. In Wenger's terms, we might suggest that **alignment** over time leads to the creation of 'normal desirable states' between teachers and students. A question then arises as to what outcomes the normal desirable state is achieving in terms of students' successful learning of mathematics and whether those involved ask this question of their practice. Recent results from the Third International Mathematics and Science Study (TIMSS) and the Programme for International Student Assessment (PISA) tests in several Western countries, including the UK and Norway (e.g., Mullis, Martin, Gonzalez, & Chrostowski, 2004; Organisation for Economic Co-operation and Development, 2004), suggest students are not achieving as well as those in other parts of the world. It seems worth inquiring into classroom norms that result in such outcomes.

Returning to Lave and notions of 'teaching as a practice', in which the knowledge of teaching is in the practice of teaching, it is possible to study aspects of this practice and provide deep accounts of both the practice and the knowledge within. The unit of analysis here is **practice**, and the focus of research is on an arena or situation which allows a study of practice—for example, a whole school or one classroom. Such a study might try to capture learning, or growth of knowledge, within teaching—teachers learning about teaching as part of their practice of teaching. This begs the question about how learning about teaching links to pupils' learning of mathematics.

The design of our study is intended to provide this link. We provided for interactivity of teachers and didacticians in workshop and school situations, in which we explored possibilities for enabling pupils to inquire into and hence learn mathematics. Fundamental to this inquiry process is that 'belonging' to the project community extends Wenger's 'alignment' to 'critical alignment' (Jaworski, 2006) through which we (teachers and didacticians) question overtly both established practices and the new approaches we design within the project. The concept of critical alignment recognises the necessity of aligning with the norms of practice in order to function within a community of practice and, as well as, or as part of this recognition, questions the ways of doing and being within the practice. In LCM, this questioning looked fundamentally at what the practice achieves or fails to achieve in terms of providing opportunity for students' learning of mathematics and considered what might offer enhanced learning opportunities. This involved questioning current practice while engaging in practice—what is possible and how—what do we want to try to achieve and what does it look like in practice? What is emergent from such engagement and questioning? We started by using 'inquiry' as a tool for learning and working toward inquiry as a **stance**—as **a way of being in practice** (Cochran-Smith & Lytle, 1999; Jaworski, 2004).

INQUIRY AS A TOOL FOR LEARNING

Within the LCM project, operationalisation of the previous ideas has involved grappling with notions of inquiry in all three layers, both separately and together. Project design had suggested that the project would be interpreted through joint activity between teachers and didacticians in workshops at the university and innovation in schools. The former should allow us to explore ideas and discuss possibilities outside the immediacy of school settings, whereas the latter would bring ideas into the practice arena with consideration of all the interpersonal and institutional factors that are part of school life. In both cases, we came to regard our interpretation of theory in practice as comprising a design/inquiry cycle of **plan, act, reflect, feedback,** which becomes a research cycle when we also **observe** and **analyse** (see Figure 15.2). The left-hand part of the figure constitutes the

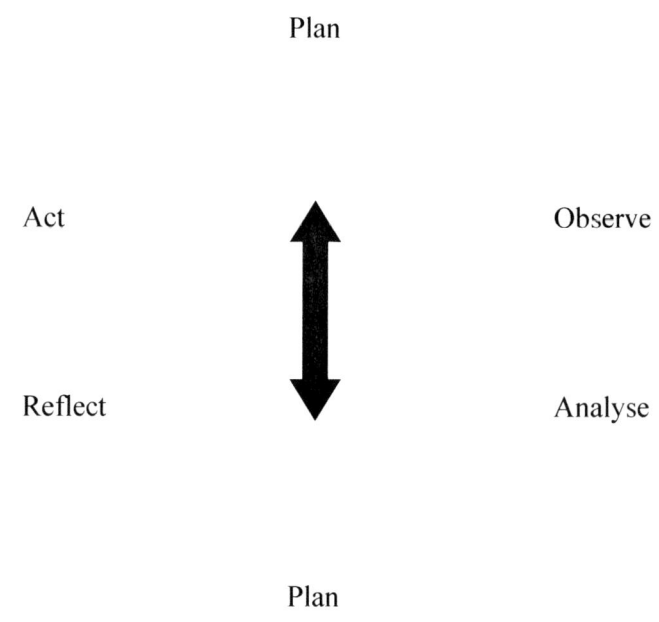

Figure 15.2 A research cycle in developmental design.

developmental cycle, **plan, act, reflect, feedback,** whereas the right-hand side constitutes the research cycle, **plan, observe, analyse, feedback.**

In the case of design of workshops, an early decision was taken by didacticians that it could be valuable for teachers and didacticians to do mathematics together in workshops as a basis for thinking about issues in learning and teaching. Thus, in advance of each workshop (we had 16 workshops during the three phases of the project), didacticians planned mathematical tasks with which all participants would engage in the workshop setting. Immediately after each workshop, the didactician team reflected on the workshop activity and outcomes, and these reflections fed back to future planning and task design. For didacticians, these actions constituted the developmental part of the activity. In addition, observation took place involving data capture in audio or video form from all activity—planning, workshop, and reflection—with subsequent analysis, resulting in a research formulation from the activity. Deliberation on possible tasks, the tasks that were designed, the overt observation, reflection, and analysis can all be regarded as inquiry tools.

In the case of activity in schools, teachers, singly or as a team, designed tasks for their classroom(s). A range of schools, from lower primary to upper secondary, and a larger overall number of people meant that design activity of teachers in schools was much more diverse than that of didacticians in planning workshops. Often school design actually started in a workshop, either with a workshop task which would be modified for use in a classroom, or with group activity in the workshop, often with teachers from several schools planning a lesson sequence. In the school setting, the tasks had to be reformulated appropriately for use in the classroom; lesson sequences had to be adjusted to fit particular school circumstances. Thus, school planning activity might involve a group of teachers or just one teacher in formulating tasks that could fit forms of activity in the particular school environment. We can see here critical alignment in practice as teachers used ideas they had generated in the workshop setting to offer opportunity to pupils while at the same time adjusting innovatory ideas to school norms and traditions. The workshop tasks, lesson sequences, deliberations within or between teachers, and adjustments for the school setting can all be regarded as inquiry tools. In the research part of the school cycle, teachers, through observation and reflection, fed back to future activity and presented outcomes in workshops; didacticians collected video data from classrooms for future analysis according to a range of research questions. A discussion of the complexities of research in the project is beyond the scope of this chapter.

The previous two paragraphs describe more overt forms of activity in the project. Less overt in some ways are the involvements of teachers in workshop planning and of didacticians in school planning. Although there were overtly many occasions when didacticians joined teachers in schools, at the invitation of the teachers, and fewer occasions when teachers joined didacticians in planning workshops (it was harder for teachers to take time out from school activity), the more hidden influences are what I address here. As teachers and didacticians interacted, either in the workshop or in school, ways of knowing and being together developed. Didacticians gleaned teachers' perceptions of their workshop experiences and desires for what workshops should offer them. Teachers perceived that didacticians sought not to give didactical instruction, but to offer opportunity to engage and to include teachers in inquiring into possibilities. As a result of such recognitions and perceptions, specific meetings were organised, some called by teachers, some by didacticians, at which views were sought and expressed, and frank exchanges allowed better understandings to grow (Goodchild & Jaworski, 2005).

ISSUES IN THE PROJECT

We have seen many positive outcomes from project activity that point to learning and development within the project. However, these matters of learning and development are far from simple. In the space available here, I focus on some of the issues that have arisen for ongoing thinking and practice.

Hierarchies in Mathematics for Teaching

Schools in the project ranged from lower primary to upper secondary. There was thus a wide range of mathematical experience among teachers in the project. Teachers with less experience expressed concern when the mathematics discussed seemed to go beyond that with which they were familiar. Conversely, teachers at the higher levels wished to work mainly with mathematics at their own level. This meant that teachers overwhelmingly preferred to work with others who taught at the same level as themselves. Didacticians felt that there was value in working together across school levels in order to develop more understanding of what was experienced by or expected of students at the different levels. The compromise was usually in favour of same-level groups, although all met together in plenary to share outcomes from group work.

Difficulties with Conceptualising Inquiry

'Inquiry' is a familiar word in the international literature in mathematics education and has represented important concepts within the LCM project as indicated earlier. Mathematical tasks in workshops have been inquiry based, and a major source of discussion has been what role inquiry-based tasks might play in classrooms. However, there is no one single word in the Norwegian language that can be used to capture the meanings associated with the word 'inquiry'. Much discussion and debate throughout this project with many groups of people has led to a range of words in Norwegian that can capture the meanings involved. Either because of, or as well as, this, it has taken time for the project community as a whole to internalise the meanings of inquiry so that it becomes recognisably a part of practice in a range of levels. During the project, the word **inquiry** has entered partially into the Norwegian language, so we find the word 'inquiry' popping up all the time in Norwegian discussion. How teachers see inquiry within their own practice has been central to research. For teachers in upper secondary schools, for example, inquiry tasks have been seen as largely separate from their curriculum-based teaching, so that, for them, day-to-day teaching cannot be contemplated as being inquiry based.

(Perceived) Conflicts with Curriculum

The teachers at the higher secondary level experienced a demanding curriculum and indicated that they could not spend time on extra activities. Their perception was that inquiry-based activities would be extra to their curriculum. So, although didacticians worked hard at producing activities which they saw as being clearly curriculum-related, teachers were reluctant to consider using such activities in their classrooms. In one case, teachers in an upper secondary school invited didacticians to

Developing Mathematics Teaching Through Collaborative Inquiry 181

work with them to design inquiry-based tasks related to a topic on linear functions. As a result of joint activity, three teachers themselves designed a sequence of four tasks which they each used with pupils, recorded on video by didacticians (Hundeland, Erfjord, Grevholm, & Breiteig, 2007). In these lessons, the teachers reported evidence of a higher level of understanding from their students than they experienced normally. However, time in planning the tasks and classroom time was greater than normal, and they felt they could not in general afford this amount of time.

Mathematics in the Classroom

Design of inquiry activity and tasks was focussed, directly or indirectly, toward pupils' mathematical development. Teachers took the results of design, from workshops or in schools, into their classrooms and invited didacticians to video record the activity. The video record and emerging analysis point to a broad range of innovative activity and pupils' engagement in inquiry in mathematics. However, analysis has focussed on the learning of teachers (and of didacticians), rather than on the learning of pupils per se, so we have no measures of pupils' mathematical development over the time of the project. For example, Jørgensen and Goodchild (2007) report collaboration and learning of a teacher and a didactician in developing activity for first-grade pupils. The video data from Jørgensen's classroom shows considerable evidence of pupils' mathematical engagement and exploration, but the paper reports the developing thinking of its authors as they designed, acted in the classroom, and reflected on their activity, rather than on pupils' learning per se. Similarly, the paper from Hundeland et al. reports from the linear functions activity in upper secondary school. Engagement of pupils in inquiry and discussion was a major feature of this activity, but analysis focuses on task design and learning of teachers and didacticians, rather than the learning per se of the pupils. This is in accord with theoretical perspectives outlined earlier, particularly addressing questions about the nature of teaching in a community of inquiry.

Tensions for Individuals and Groups Working between Project Aims and School Traditions

Schools in Norway up to the age of 16 organise pupils in undifferentiated class or year groups. It is illegal to set or stream. Teachers form year teams, so that the teachers working with one year group work and plan together for students in their year. However, the LCM team in a school often crossed several year teams. This made it difficult for teachers in the LCM team to meet each other during school hours and to plan for collaborative work. It was correspondingly difficult for didacticians to meet with a whole school team in school hours. Difficulties in meeting together were given as a reason why some teachers planned and innovated singly with their own class, rather than through working together in their project team.

TENSION AS A SOURCE OF LEARNING: EMERGENT VERSUS ESTABLISHED ACTIVITY

Issues exemplified earlier have arisen from tensions between the project community and established communities. These tensions can be related to questions of **ownership** within the project. The project originated through a proposal by didacticians to the research council and became a reality when the research council funded the project. Didacticians designed the project, establishing aims and approaches. Thus, workshops, their aims and activity, ideals for school activity, and so on were introduced by didacticians. Schools were invited by didacticians and responded voluntarily to participate. Contracts were signed agreeing to certain forms of activity and input (for further detail, see Jaworski, 2005). At every stage of the project, didacticians sought the views of teachers and discussed with them where possible the best lines of action. Nevertheless, and unsurprisingly, in the first phase, many teachers gave the impression that they were responding to the wishes of didacticians and wondering what the benefits were for themselves. In the second phase, when it was clear that teachers' views were sought and workshops were planned according to teachers' wishes, some balance was achieved; in Phase 3, there seemed to be a greater extent of participation by many teachers. Nevertheless, in the LCM project, partnership was skewed toward the ideas of didacticians in both design and implementation.

These tensions and their resolution have been important to the project. They have been brought to light by the recognition of issues emerging through project implementation. Resolution has often involved compromise with implementations of design and to some extent with the aims of the project. Such compromise has arisen mainly from factors in established communities: for example, the group structure of the school system, perceptions of curriculum demands, or teachers' feelings about mathematical knowledge. In the power imbalance, planning for activity came largely from didacticians and drew on didacticians' knowledge of research literature and theories of learning—part of the established structures of university life. As the project worked through its three phases, forms of practice and ways of dealing with issues and tensions emerged. In terms of a research cycle at the project level, we can see compromises being recognised, reflected on, and absorbed into project culture. Both teachers and didacticians were brought up against issues in their own practices and expectations, and they had to rethink their position, reformulate ideas, and reconsider roles within the project (e.g., Cestari, Daland, Eriksen, & Jaworski, 2006; Daland, 2007). This pushed them into a position of addressing aspects of their own familiar practices through new lenses, taking a critical stance and seeking outcomes favourable to their aims within the project.

Shifting awareness with respect to such critical alignment can be exemplified by the words of one teacher who, in a focus group interview at the end of Phase 2, spoke as follows:

> *Agnes:* In the beginning I struggled, had a bit of a problem with this because then I thought very much about you should come and tell us how we should run the mathematics teaching. This was how I thought, you are the great teachers. [. . .] but now I see that my view has gradually changed because I see that you are participants in this as much as we are even though it is you that organize. Nevertheless I experience that you are participating and are just as interested as we are to solve the tasks on our level and find possibilities, find tasks, that may be appropriate for the pupils, and that I think is very nice. So I have changed my view during this time. (Translated from the Norwegian by Espen Daland; Daland, 2007)

During the third (and final) phase of LCM, funding was granted for a second project to build on the processes and practices of LCM. The subsequent project, Teaching Better Mathematics (TBM), began with a more equitable balance of power. Both schools and university have their own funding, and activity is planned by a committee representing both partners. This illustrates one important outcome of LCM in organising developmental research with respect to systemic factors and equity issues.

LEARNING ABOUT TEACHING AND LEARNING—GROWTH AND IDENTITY

Analysis shows considerable evidence of classroom experimentation, of teachers and didacticians learning from their project activity, and of changes in perceptions of what matters for mathematics teaching and learning. A range of publications charts this learning (http://fag.hia.no/lcm/papers.htm). What this means varies for the people involved according to their institutional setting, their own roles within the setting, and their responses to the project. Characterisations have been offered in a book (with some chapters in Norwegian and some in English) which has emerged from the project to which all didacticians and most school teams have contributed (Jaworski et al., 2007). Doctoral students, who at the time of writing this chapter are in the last stages of thesis work, are employed also: in two cases in the university as didacticians, and in one case each in a higher secondary school and in a local professional development team. Three of the original eight schools (primary and lower secondary) are now part of the TBM project, with the upper secondary schools indicating their wish to join. In the TBM project, teachers from LCM schools are taking leadership roles in bringing new schools into developmental activity. Members of the LCM project, didacticians, and teachers, at the invitation of the research council, led a national conference to which teachers and didacticians throughout Norway were invited. These events and outcomes show evidence of project-related identity building, both for individuals and groups and associated agency in developmental practice.

NOTES

1. The LCM project was supported within the KUL Programme (Kunskap, Utdanning og Laering [Knowledge, Education and Learning]) of the Norwegian Research Council (Norges Forskningsraad [NFR]), Project No. 157949/S20.

16 What Does a Discourse-Oriented Examination Have to Offer Teacher Development?
The Problem With Primary Mathematics Teachers

Tansy Hardy

This chapter maps the question in the title into the context of teachers of mathematics and their professional development.

The language (in its broadest sense) that circulates around the professional practice and development of teachers constitutes a discourse-rich field. This discursive world includes not only what infuses mathematics classroom environments and more general education discourse, but also media representations together with public and personal understandings from wider aspects of life. All are present, even unavoidable, and have significant effects on our knowledge and understanding of teachers' development.

My discussion starts by showing how 'getting to grips with the ways in which language is entangled with reality' (MacLure, 2003, p. 5) offers productive ways of examining teacher development in current times. In this approach, discursive resources are used to illustrate what is brought into operation through current language of mathematics teaching and to examine the effects of mathematics education discourse on the practices and forming identities of teachers and learners.

FADS AND FASHIONS

A discourse-oriented approach looks at how terms take on their meaning. The following extract from a radio news interview[1] with the shadow education minister provides an example. Discussing recent education reforms, he said, 'There have been a lot of progressive fads in teaching approaches. Now it's time we stuck to traditional approaches that have been shown to work'.

A private school headteacher responded, 'We don't have to follow the latest government fashion. We stick to what we know works'.

These comments may seem unremarkable. They draw on familiar language well used by politicians. However, a discursive approach prompts us to be suspicious of the effects of the unremarkable: *'suspend your belief in the innocence of words and the transparency of language as a window on an objective graspable reality*' (MacLure, 2003, p. 12, italics original).

Rather than asking what the best/correct/recommended teaching approach is, we are prompted to consider what knowledge is produced and how possible meanings might become limited or fixed. As this 'knowledge-making' process is brought into operation through discourse, a scrutiny of the play of terms in this dialogue can show the process through which knowledge is brought into being. This illustrates how the use of particular terms has an effect on 'what we know' about teaching approaches.

'Fads' and 'fashions' come together with 'what we know works' as these terms take on their meaning and produce knowledge about 'teaching approaches'. We come to know that some teaching approaches 'work'; these are in opposition to fashionable teaching approaches, and such fads will pass. Schools and teachers who use these approaches are also produced as subject to government fads or as resiliently sticking to what works.

This type of scrutiny forms part of a toolkit of strategies that can be assembled to expose how discourses and their associated practices produce familiar descriptions and 'truths' about mathematics teachers and their professional identities. In the rest of this discussion, I want to demonstrate how this knowledge-making produces complex statements that are not straightforward to unpick or challenge and, as such, can set up unquestionable truths.

The discursive turn I propose allows an opening-up—a disruption of such 'truths' in a particular site or enterprise. This is paralleled by the shift in research questions advocated by Heather Mendick et al. in Chapter 7 (this volume).

BEING SUSPICIOUS OF 'TRUTHS'

The discursive world refers not just to language, but more broadly to:

> the multiplicity of meanings that attach to (and divide) the people, spaces, objects and furniture that comprise its focus—the teachers, children, classrooms, textbooks, policy documents and to the passion and the politics that are inevitably woven into these meanings. (MacLure, 2003, p. 12)

In this sense, the field of teacher development in England is a discourse-rich one. There is a strong voice of policy and government initiatives. Discursive tools include assessment instruments, evaluations, profiles, and target setting. For both initial and in-service teacher education, there are taxonomies of teaching standards and competences to be evidenced through portfolios and learning logs. Multimedia guidance on teaching approaches abounds, and there is daily representation in news reporting. These are brought together in professional discourse to create categories and truths about teachers and teaching (e.g., 'effective' teacher, 'good' learner, teaching strategies 'that work') and are in play in any understanding of the latest government initiatives. This is a text- and image-saturated field.

The context of national education policy initiatives to improve mathematics teaching offers a demonstration of the value of a discourse-oriented approach to research in this field. These initiatives address a repeated need for improvement, through tackling so-called obstacles to improving children's learning. Their projects attempt to define and explain the process of teaching in order to capture the essence of the field. They are premised on a need to scientifically analyse, list, and categorise what determines effective mathematics teaching. Such projects seem to sustain a perpetual reinvention of 'what works' (although the partiality of any account of 'what works' is not acknowledged). Even those in the early stages of their teaching career are well aware of government's dissatisfaction with the status quo and this undeniable call for reinvention. Associated 'retraining' is often experienced by teachers as 'initiative overload' (Smithers, 2003), the linked guidance as advice overload. There is too much to take onboard; something has to be ignored, and inconsistencies have to be overlooked.

However, the repeated and chronic failure of initiatives to give the desired improvement is hard to give voice to. This is an effect of the discursive practice—fixing knowledge and understandings in a way that is surprisingly hard to open up. The discursive strategies I set out follow an injunction to take the workings of discourse seriously and to study the way it constitutes 'regimes' of truth that close off alternative accounts (Foucault, 1972). They offer some access to how systems of governance, rather than reflecting innocently the world of teaching, actually create knowledge about the area—knowledge that is inevitably partial.

The rest of this discussion is set in the context of primary mathematics. As the processes indicated organise this field, producing available meanings and knowledge, the subjects—'the learner', 'the teacher', and 'mathematics'—are inscribed within the language and practices of this education discourse and are produced as determined positions. Such positions are produced through a double-sided process where people can function as agents, active in constructing their professional selves. However, the way people act is linked to a way of thinking derived within discourse practices. These guide and can restrict how people think about themselves. (Subsequent meanings and consequences of any action are not totally determinable.)

To exemplify this feature, I have extracted a short section from a course document from initial teacher education—a particularly pertinent site. These learners are positioned in transition within overlapping discourses. In their professional learning, their own previous learning experiences are present. They are also engaged in developing an image of their future professional selves.

The document gives strategies for how to raise the standard of placement grades on the course—that is, how students could achieve higher grades on school placements (a concern not least for future course inspections).

The following guidance was given:

In English: ensure better links between school and university practice (e.g., more use of drama and children's own experiences and ideas).
In Maths: more structured planning (e.g., give students specific scenarios to plan for, such as a group/child is having difficulty with understanding aspects of mathematics, a child/group is having difficulty counting in 10s crossing the 100 barrier; plan a series of lessons which break down the learning objective into smaller learning steps).

Discrepancies between the guidance for these two subjects are striking. For English, it takes only two lines to say what is needed. The guidance is in terms of general ideas, such as 'better links' between what the students do in schools and what they are introduced to in university. This linking can be achieved through addressing broad and creative areas: 'use more drama and use children's own ideas'. There is space for students to write themselves into this subject.

In maths, the guidance is in terms of more structuring. There are difficulties to overcome here. It is already 'known' within the profession that maths is a difficult subject, one where children have difficulties. Students may know this from their own learning experiences. Further, in this subject, planning is difficult and remedied by more structure. This is reinforced with a specific, detailed example to help tutors and students see how to improve. The example is technical, reinforcing the mantra 'break the maths down into small steps'.

When the subject is mathematics, the guidance takes on a technical, specific form, promising a solution for children's, students', and tutors' difficulties. With English, there is a very different formulation. From course documents such as this, beginning teachers learn that, unlike English, for maths they have little autonomy and little to offer in creating rich learning experiences for children. They must follow a technical step-by-step approach.

This illustrates the production and operation of 'truths' that I examine further in the next section.

THE PROBLEM WITH PRIMARY TEACHERS' MATHEMATICS

Mathematics teaching in primary schools is portrayed in the last example in ambiguous terms. The prevalent theme of difficulties with this teaching is repeatedly reported in media stories and research findings. In this section, I collate some extracts from recent research reports, news reports, and other publications. Together they illustrate how the question 'What's the problem with primary students' mathematics teaching?' can become 'What's the problem with primary students' mathematics?'—in what I argue is a faintly noticed essentialising shift.

Many professional development programmes take for granted that courses should be designed to increase primary teachers' understanding of mathematics as well as develop effective practices. Documentation commonly portrays a 'problem' with the students' personal maths knowledge. This 'produces' a teacher whose mathematics knowledge is the problem: deficit or flawed in some way. The introduction to a recent preservice course textbook, *Mathematical Knowledge for Primary Teachers*, reads: 'One of the problems to be overcome by many seeking to teach mathematics to young children is that they have the wrong kind of understanding of their subject' (Suggate, Davies, & Goulding, 2006, preface).

A renewed concern with teachers' mathematics knowledge is notable in government strategy. For example, in this inspection report, schools' uncertainty in putting national strategies into practice is associated with a lack in what mathematics teachers know:

> Findings suggest that although schools support the government's vision [. . .] few have confidently embraced the PNS [Primary National Strategy] through major restructuring of [. . .] the way in which learning is organised.
>
> Inspectors found that although teaching in English and mathematics was improving there remain concerns that some teachers still do not have the subject knowledge they need to raise attainment further in English and mathematics. (Ofsted, 2005)

This report positions teachers' mathematics knowledge as an obstacle to improvement in children's learning. Its findings are premised on the effectiveness of wide-scale systematic change in improving the standard of children's attainment. This premise is questionable. For example, a 5-year study into primary mathematics in England finds:

> Finally the results highlight the complex yet weak relationships between teaching and learning, and in particular that a major attempt at systemic change has had at most a small effect on attainment in most areas of numeracy. (Margaret Brown et al., 2003, p. 22)

WHO CAN BE BLAMED?

Textual resources work together in the discursive field to carve up primary maths teaching and to open up new forms of blame. The blame slides from an initiative and its promise—the National Numeracy Strategy and its framework—to the way teachers put this into practice by shifting attention for the 'difficulty' with policy to difficulties with practice, and further from lesson structure to the teacher her or himself. The problem is no longer attributable to the guidance, but to how this is put into practice. The recommended teaching approach remains unquestioned. The problem slides again from practice to teacher and to teacher's mathematics knowledge.

We come to know that the problem with primary mathematics teaching is with the subject knowledge of those who are trained to teach mathematics in primary schools. This plays out in the UK context as a need for entry audits and exit numeracy tests for initial teacher training (ITE) students. Examining, measuring, categorising, and qualifying are the tools of surveillance that regulate and classify these student teachers. These tools also reproduce our knowledge that their knowledge of mathematics is insufficient. The truth of this holds sway despite well-recognised findings that formal qualifications are not reliable indicators of effective mathematics teaching in the primary years (Askew et al., 1997).

LOCAL AND NATIONAL PROJECTS

Strategies to redress the lack in teachers' mathematics and research into this 'problem' give the focus for my next examples. I draw from a range of reports to offer statements that I believe encapsulate 'what we think we know about' primary teaching. (I do not suggest intended support on the part of the authors of these extracts for unquestioned truths about primary maths teaching, nor do I exclude myself from those caught up in this inevitable working of discourse.) Some of the extracts are from others who alert us to the need to question common understandings and to look for alternative ways of framing the field. Through my strategic examination of how these extracts 'work' together, I suggest some current 'truths' which can be interrogated to offer an alternative reading. My interpretations offer possible readings. This follows the spirit of a discourse-oriented enterprise which engages with the provisionality running through this approach and the similar nature of what might be called its findings.

There are many local initiatives engaging ITE students and teachers in what are termed 'richer', more interactive, open-ended mathematics teaching. Each local version may be distinctive in its format and premise. It is, however, striking how often these projects report some difficulties with teachers' mathematics subject knowledge.

To set the background for a longitudinal study of professional development to build primary teacher expertise in teaching mathematics, the authors remind us that it is known that:

> Mathematics education poses substantial challenges for elementary teachers, who often lack the knowledge of mathematics required to effectively implement reform-oriented mathematics programs (Ball, 1988,

1990; Ball, Lubienski, & Mewborn, 2001). Many elementary classroom teachers have difficulty providing rich mathematical experiences, recognizing the mathematical connections that children are making [...] A teacher's weak understanding of math may prevent him/her from recognizing and furthering the important concepts that are inherent in mathematical activity. (Suurtamm & Vezina, 2004, p. 1)

In the article 'Does It Matter? Primary Teacher Trainees' Subject Knowledge in Mathematics', ,Goulding, Rowland, and Barber (2002) report on research in their local contexts:

> The research identified weaknesses in understanding, particularly in the syntactic elements of mathematics, and a link between insecure subject knowledge and poor planning and teaching. The dilemmas and policy issues which this focus on subject knowledge presents are discussed. (p. 689)

These two extracts point to the difficulties teachers have with teaching mathematics. Now primary teachers not only do not know enough maths, but also do not have the required sort of knowledge. They lack the forms of mathematical understanding needed to develop effective pedagogic strategies. Their knowledge is blind to mathematics' modes of inquiry and the integrity or 'connectedness' of its content. The association of 'secure' knowledge of mathematics (McNamara et al., 2002) with effective pedagogy extends the scope of the constructed problem.

The description of valid mathematics knowledge as 'secure' generates the possibility of teacher's knowledge as insecure and evokes a teacher who is insecure. Regardless of further training or 'top up' undertaken by teachers, they always emerge with the wrong sort of knowledge. By association, their teaching and planning become suspect, an aspect of their professional selves that can become categorised as insecure. Through slight references to affective and inner states, possible positionings and identities are produced for teachers to take on:

> Essentialising forms of [psychology] are those types of [psychology] which generate internal categories of personhood that are unchanging and timeless, that come to be inescapable, and hence that bear a determining influence of sorts on the person in question. [...] one effect of a strong (essentialising) discourse is the production of a fixed fictionalised identity.
>
> Why does this matter? This discourse determines in so much as that person comes to understand themselves and how they understand others. (Parker, 2004, p. 139)

This trick of knowledge/power (Foucault, 1972) masquerading as commonsense leaves us unaware of the effects of our practices on ourselves and others. The truths produced can speak of deep internal states of being, to our sense of ourselves and others. This takes hold to generate what can be termed 'obedience' as teachers take on such positionings to understand themselves and their teaching. Pignatelli (1993) describes this in his work on teacher agency: 'The obedient subject is produced and sustained by a power faintly noticed and difficult to expose; a power that circulates through these small techniques among a network of social institutions such as the school' (p. 420).

This can be seen in Bibby's (1999) reporting that there is 'discomfort' for some teachers with their own level of mathematics content knowledge.

References to internal states (confidence, insecurity) in the explanation of the problem with primary mathematics teaching can essentialise teachers in the way that Ian Parker articulates. This implicates not only their knowledge and competence, but also establishes a more personalised failing at an affective level. This expands the field of influence and regulation of the discourse. What we know of the problem and its explanation reemerges as a truth: that 'teachers are not confident with mathematics and not confident with teaching mathematics'. In the next section, I examine how this 'finding' functions as a truth and some of the ways in which it goes unquestioned and unquestionable.

It is remarkable how often the word 'confidence' occurs in reviews of what we know about primary teachers and their relation to mathematics and its teaching. In a professional development video, Anita Straker features in the commentary and summarises: 'The positive outcome of the project is the teachers are much more confident and the children are much more confident' (Hamilton Maths Project/National Numeracy Project, 1998). A brief search in the maths education arena will also reveal a proliferation of occurrences where confidence and competence go hand in hand (e.g., in the title of the review of research by Kyriacou & Goulding, 2004). The entangling of performance/attainment in mathematics with confidence appears to be taken as commonsense, and the terms are conflated in an almost unquestioned presumption that confidence is a necessary partner to competence.

To give two further instances: On a government website, study materials on their knowledge of mathematics are introduced to teachers:

> In order to gain confidence and competence in mathematics subject knowledge, you could audit your knowledge [. . .]
> Ask yourself, are you: familiar with this? able to do some questions involving it? confident that you could explain it to someone else? aware of pupils' difficulties in this area?
>
> These questions will give you an indication of your competence in the area. (Teachernet, 2007)

Similarly, the UK's virtual National Centre for Excellence in the Teaching of Mathematics offers:

> Mathematics subject knowledge audits:
>
> Are you looking to develop your subject knowledge, skills and understanding?
>
> Do you want to feel more confident in knowing what you know?
>
> Do you just enjoy doing mathematical questions and quizzes to pass time? (National Centre for Excellence in the Teaching of Mathematics, 2007)

Discourse-oriented research attends to such reoccurences. The next example, from my own research activities, uses the notion of 'confidence' in learning as a tactic to interrogate truths about preservice students' relationships to mathematics and mathematics teaching. A pertinent site: It is the confidence of these students that is often described as the problem with primary maths teaching.

This project worked with a group of students who had chosen to take mathematics as a specialism within a primary education course. This involves them in learning some mathematics at their own level. The course has been described as an opportunity to learn to learn mathematics, to develop 'secure' mathematics knowledge, and to renegotiate their relationship with learning and so teaching mathematics. In individual and group interviews, students were asked to complete some unfinished statements about learning mathematics and what they thought confident learners were like. My students had no difficulty identifying who is confident—that is, other students that they thought were confident in maths.

A sample of their remarks follows:

> For their course group the confident learners: respond better to being questioned; they are the ones who always volunteer; they'll try for answers even if they are not sure if it's right; they are less likely to be embarrassed by their mistakes; they can explain things to others; they are more willing to seek help than those lacking in confidence
>
> Of themselves: I will have a go in maths when I know the subject very well; when I'm 99% sure; if I can have a go on my own; if no one is watching if I get it wrong.

Themes of 'speaking out; offering answers; explaining; asking for help; taking risks; having a go' recurred. These refer to what they and others do when they are learning. In my analysis (Hardy, 2007), I found that those I asked had no difficulty identifying forms of participation that indicated confidence in their peers. The attribution of confidence seems to follow from predominantly performance-based elements. Confidence is performed. Only one student suggested that it was possible to be confident and not prepared to speak out.

The (overly) easy way that learners are labelled as confident or not from what they do in learning situations was also noted in Burton's (2004b) work:

> They (teachers) had no difficulty identifying confident students. They spoke about willingness. In their terms, willingness to have a 'go', ask questions and challenge them was evidence of confidence. Students wanted an atmosphere where you were not afraid to make errors. (p. 363)

Confidence appears as a faintly noticed modifier, produced as an object, a category—a confident learner, a good learner, good at maths. It emerges as an essential characteristic, an inner state. Its potency is in its 'hard-to-define' nature. When described, it is in terms of forms of participation and performance, but it is attributed as an inner state: a thorough conflation of competence and confidence. This smoothing over works to deflect any questions of alternative descriptions that could be productive.

To dislodge this 'closing down', a discursive approach looks for oppositions, repetitions, tensions, and inconsistencies. In my data, a discrepancy can be seen between students' articulations of conditions when they would contribute and their descriptions of the acts of their peers to whom they would attribute confidence. An erasure appears—a need to stay unaware. Foucault (1972) expresses this as the will not to know where discursive disciplines exist precisely in order not to know things.

Throughout this discussion, I have offered examples that reveal the 'truths' that masquerade as obvious and exemplified the processes through which these operate as unquestionable. This 'laying bare' is a start to opening these to challenge and to reformatting. As MacLure (2003) summarises,

> A discourse-based educational research would set itself the work of taking that which offers itself as common-sensical, obvious, natural, given or unquestionable, and trying to unravel it a bit—to open it up to further questioning. (p. 9)

My concluding remarks attempt to take seriously what this might mean for my practice in the field of mathematics teacher development.

CONCLUDING REMARKS

Another iteration of 'solving the problem with primary mathematics teaching' emerged in the UK in 2007. The Williams Review was set up with the remit to consider and make recommendations in areas of primary mathematics teaching 'known to be' key to improving children's attainment in numeracy (http://www.standards.dcsf.gov.uk/primary/mathematicsreview).

The first two of these areas:

> What is the most effective pedagogy of maths teaching in primary schools and early years' settings?
>
> What conceptual and subject knowledge of mathematics should be expected of primary school teachers and early years' practitioners, and how should initial teaching training and continuing professional development be improved to secure that knowledge?

On first reading, I suspect that the broad areas identified by these two questions will appear appropriate and 'commonsense'. Again, the areas under my interrogation are sustained as areas of difficulty for which fixes need to be sought. This is a pertinent reminder of how easy it is to be caught up in treating this knowledge as unproblematic and unquestionable—a reminder voiced by others, for example:

> Agencies of power sustain control not by repression but by eliciting consent. People in modern institutions are conditioned to accept being an object to others and a subject to themselves. The very processes we use to inscribe our self to our self put us at the disposition of others. The task of creating rational, autonomous persons falls initially to pedagogical institutions. (J. Roth, 1992, p. 691)

From this I propose that the field of teacher development in mathematics needs to entertain the possibility that it is the very conceptualising of primary teachers' professional knowledge of maths which, in its articulation, generates and condemns teachers to having faulty knowledge. That is, it is the attempt at better description (in order to better understand) that produces the problem. By offering itself as an explanation (e.g., that confidence is lacking) of what it purports to describe, discourse smoothes over inconsistencies, discharges any challenge to assumptions, and closes down alternatives.

Some in this field have seen the need to ask how we can develop clearer theoretical and practical understandings of 'secure' mathematical knowledge and its relationship to pedagogy. In their Research Monograph McNamara and colleagues (2002) pose this in order to inform further questions about ways teachers can develop 'secure' forms of mathematics knowledge through their training and subsequent practice in teaching.

This call for a wider gaze may be motivated by the desire for a more inclusive and more relevant understanding of the relationship between teachers' mathematics knowledge and their mathematics teaching. However, with a discursive orientation, it is important to be aware that emancipatory projects take place within the discourse practices they seek to disrupt and are never fully free of their regulatory effects.

A call for clearer theorising could equally well emerge from a desire for a more comprehensive description and categorisation of 'teaching mathematics' (i.e., a belief in the need for an overarching theorisation of the subject and all who are implicated in it). Whatever the intent, its practical effect may be to leave the field susceptible to more comprehensive regulation, allowing less space for resistance or deviation. In order to make a claim that things are other than currently described, one is driven to articulate in the very language that is being criticised. Alternatives become hard to consider. In other words, language 'cleverly cannibalises' liberal humanist terms in making what appear to be self-evidently sensible statements (Davies, 2003, p. 98).

This leads to a question about what form of reconceptualisation could open up alternative understandings. An opportunity is offered by the incomplete nature of current models of the knowledge required for teaching mathematics.

The theoretical and practical implications of what is asked, and how, have to be acknowledged critically, as does, importantly, what is on the periphery of the zoom of the research lens (Adler & Lerman, 2001; cited in McNamara et al., 2002).

One tactic would be to work specifically with the periphery. In relation to theoretical models, 'part of their usefulness can be at their borders, in what they specifically exclude and include in particular instances of their use' (Corbin, 2000; cited in McNamara et al., 2002, p. 66).

Another strategy available for the interruptive project of discourse analysis is to ask different questions in radically different terms. Placing this back in the field of teacher education, I can ask what, as a teacher educator, I might do differently tomorrow as I work with teachers, design courses, plan activities, and configure relationships and interactions. How might I describe these in new ways as I report my research? Several of the courses that I have referenced earlier clearly intend to offer new forms of engagement with mathematics. Heywood (2007) reports on a course for preservice science teachers in a subject field which is also experienced as strongly regulated:

> problematizing subject knowledge through direct experience of learning in science, particularly in those areas that are known to be difficult, constitutes a productive way of turning a deficit model of teacher subject knowledge into a positive experience with considerable potential for the development of pedagogy.

His example re-stories course activities. Subject knowledge is not presented as an object, as knowledge to be learned, refreshed, or topped up. Rather, it is presented as something to be challenged and interrogated. He has interrupted how the course talks about itself and so the relationship that students form with it and their teaching selves.

Another example of a question that would open up new possibilities is given again by McNamara and colleagues (2002):

> How can students develop a capability for working on their own professional development (in mathematics education) in a way that relates to their personal aspirations of what it is to be a teacher? (p. 33)

This offers a new lens to examine my own practice. In the mathematics course I discussed earlier, I offered activities that professed to allow students to connect their aspirations of how they want to work as teachers with their learning in maths. Rather than focussing on the majority response when evaluating the course's effects, I can cast my gaze to the periphery. From this boundary, I am already aware of an interesting tension. Some students appear to struggle with open investigative mathematics for themselves and yet are clear that this is something they wish to include in their work with children. Although for most students the experience of learning on the course is reported in positive terms, for some this seems disconnected from their own aspirations. The course and the students are in these terms partial failures. For future research, further boundary work could illuminate how subject knowledge may emerge in different terms and how new positions and relations become available.

NOTES

1. BBC Radio. 4 News interview on 16/5/2007.

Part VI
Endings

17 Identity in Mathematics
Perspectives on Identity, Relationships, and Participation

Patricia George

INTRODUCTION: IDEAS OF IDENTITY

Mathematical identities have largely been cast in this book in terms of the relationships people form with mathematics and, importantly, with other actors and processes in the environs of mathematics. Given the educational focus, a main (but not sole) concern is with the mathematical identities of students, but also with teachers and trainee teachers as the main mediators of student learning who bring their own mathematical identities with them to their classrooms. There are concerns with the remnants of earlier (and ongoing) mathematical experiences on adults, including those who have, for a variety of reasons, come back to mathematics, and how these experiences shape their present-day processes of becoming. Varying conceptions of identity have been used to explore these relationships, falling into three broad groupings: sociocultural, discursive, and psychoanalytic. In this response, I draw across the chapters and approaches, but primarily on a sociocultural perspective to question whether mathematical identities matter.

All three perspectives understand learning and identity as related and thus present a social conception of learning. In particular, sociocultural theorists Lave and Wenger (1991; Wenger, 1998) see the process of learning as identity-in-the-making. Holland et al. (1998) develop these ideas, broadly defining identity as people's self-understandings formed as a byproduct of their lived experiences in particular cultural worlds and social activity within those worlds. Thus formed, identities are 'hard-won standpoints' (p. 4) from which individuals form self-understandings of and for themselves. In this volume, the cultural world in question may be seen as the activities of teaching and learning mathematics in formal places, including mathematics classrooms in schools and colleges, lecture and seminar rooms in universities, and a range of adult education settlings. Holland et al. (1998) note that identities do not 'arrive in persons in their immediate social milieu already formed [but . . .] happen in social practice' (p. 7), so that identity formation is an ongoing, ever-evolving process. This makes the idea of participation and the nature of that participation paramount. In relation to this,

Holland et al. (1998) have suggested that not all people develop 'much of 'an' identity' (p. 190) in particular cultural worlds as they may not ever be 'sufficiently engaged' by that world. However, the chapters of this book have not addressed this notion—that a person may develop not much of a mathematical identity, or a weak identity, as a byproduct of experiences of learning mathematics. Indeed, the mathematical identities highlighted in the pages are strong, whether positive, negative, or amalgams of the two.

The response I present here addresses this issue. In the next section, I explore the key ideas in the title of the book of relationships and participation through the five strands of assessment and selection, curriculum, pedagogy, teacher development, and choice. This initial section is an assemblage of ideas, thoughts, and questions that have come from reading the chapters, as well as attending the seminar series of which the book is a culmination. Emerging from this assemblage is my questioning, in the following section, of whether mathematical identities are always strong and central enough to merit the kind of attention they get in this book. I suggest that in fact they are, and I explore why it has become crucial to consider identities in mathematics education. I conclude by teasing out some of the implications of learning as identity in relation to mathematics.

MATHEMATICAL IDENTITIES: EXTENDING IDEAS OF RELATIONSHIP AND PARTICIPATION

This book shows that many 'things' mediate the mathematics that people experience and that gets taught. The strands chosen also serve to emphasise that, for the most part, the mathematics in play here is **school** mathematics. The reference to school goes beyond the boundaries of primary and secondary education and includes how wide ranges of educational institutions formalise, structure, and package mathematics for teaching and learning. Accordingly, the book focuses mainly, but not exclusively, on a more institutionalised 'type' of mathematics; the possibilities within this context are different from other more every-day forms—for example, street mathematics (Nunes, Schliemann, & Carraher, 1993), where different identities would inarguably be developed because, in this different cultural world, learners would participate differently and experience different relationships with mathematics. Mathematics is imbued with power, and the mathematics in educational institutions particularly so. As is illustrated in what follows, mathematics in such institutions is characterised by cultural practices and other 'things' that come to form part of how mathematics is experienced and viewed. These 'things' are less evident in the informal strategies that constitute people's day-to-day and out-of-school use of mathematics in problem solving, and there is thus a greater sense of ownership or control of the mathematics in this latter context.

Assessment and Selection

Assessment and selection procedures are important carriers of messages which frequently go beyond the objectives of teachers. This is particularly important for mathematics because, as Jeremy Hodgen and Rachel Marks' chapter shows (see Chapter 4), certain selection processes, notably ability grouping, are carried out in mathematics more than in other subject areas and at earlier and earlier stages of a student's life, supposedly based on assessment results. The lived reality of this practice is that younger and younger students are developing (and having to do so) self-understandings specifically for mathematics. Such early experiences can be poignant, coming to constitute defining moments of an individual's life and directing that life's trajectory. Mark Boylan and Hilary Povey's account of Louise (see Chapter 5) provides a good example here, where the adult Louise refers to being age 9 again whenever she has to deal with mathematics. As noted by Lerman (2001):

> given the age range covered by compulsory schooling, participants' identities are at their most formative, and children are particularly vulnerable to the regulating effects of social practices . . . the school classroom is particularly affected . . . since [practices] are often of greater significance to the students than the intentions of the school and the teacher. (p. 99)

If one accepts that ability grouping at earlier stages has been an increasing trend in mathematics education, then there are implications for curriculum, pedagogy, and teacher development processes.

Curriculum

Curriculum regulates what mathematics students are exposed to and so constrains the mathematical identities they develop. Via the curriculum, pedagogical practices of teachers are managed, different types of mathematics are dispensed to supposedly different sets of students, and students are differently positioned to participate in and experience different types of mathematics. Messages from the curriculum would suggest an **expectation** for these different sets of students to participate in (i.e., do) mathematics differently, as is clear in Candia Morgan's chapter (see Chapter 9). Consequently, students experience different relationships with mathematics as the nature of their participation in the mathematics is structured to be different. As Birgit Pepin (Chapter 10) shows, these curriculum differences are importantly mediated by teaching practices. Her analysis of the use of the mathematics textbook as an important curriculum material illustrates how these teaching practices are grounded in differing cultural discourses. In the UK, for example, a whole industry has been built around the powerful discourse of 'ability' so that different versions of the same mathematics textbook are used for students at the same grade level, but perceived to be of different 'abilities'. The use of these texts is further mediated by teachers at the classroom level, to include the practice of not using a text at all—seemingly in efforts to 'protect' students from the mathematics content of the texts. However, as Birgit notes, in France, all students are exposed to the same mathematics texts on the basis of an educational discourse that stresses equality of entitlement to the curriculum.

Pedagogy

Identity as relationship when associated with mathematics underlines how a person/student forms a relationship not only with the mathematics, but also with the environs of the mathematics to include the teacher, their pedagogic methods/practices, the teacher's own mathematical identity on show, the social circumstance of the learning, the textbook and the curriculum, as well as other nonschool factors. These are connections that are unwittingly yet necessarily made when a student learns mathematics. For example, in Tamara Bibby's chapter (see Chapter 11), students form relationships with their mathematics teachers, even with Miss South, where her pedagogic actions would seem to deny the possibility of such connections being made. The children interviewed in this chapter show that for them, to talk about a subject is to talk about the teacher of that subject, illustrating the importance of the teacher and his/her consideration of them in their own learning processes. The considerations the students value do not necessarily reside within the contents of the subject matter. The idea of relationship suggests some giving and taking amongst the actors involved, as well as some compromises. It is, however, easy not to see (school) mathematics in this light—that is, as disposed to allowing compromises, with its perceived nature as hard and unyielding. It is also instructive that the pedagogical actions of some teachers of mathematics work to re-enforce this stereotype of mathematics as hard and unyielding. Within this same chapter, the relationships that students develop with mathematics based on their experiences with Mr. Blake and Miss South provide contrasting examples of teacher influence on students' developing mathematics identities, and arguably these influences have more to do with the pedagogy of the teacher than with mathematics per se.

Teacher Development

Teachers thus play a crucial role in how students come to know mathematics and, by extension, the relationships students may then form with the subject. Teacher development processes have become crucial arenas for bringing awareness to and addressing issues of identity, as Barbara Jaworski (Chapter 15) illustrates. Like Tansy Hardy (Chapter 16) and Pat Drake

(Chapter 14), she does not look only at the mediating role of mathematics teachers, but also at the mathematical identities that teachers bring with them to the teaching and learning of mathematics. In addition, they highlight a need for teachers to become more aware of the multiplicity of identities that students bring with them to mathematics learning, and how mathematics teaching may be structured so as not to deny voice to these other aspects of students' identities. Yet a paradox exists here: although research in education has highlighted this need for teachers to become more aware of this multiplicity, Tansy shows how government and other rhetoric may increasingly be funnelling preservice teachers to adopt more restricted preset positional identities for themselves and for mathematics teaching. Although Pat provides useful illustrations of how some teachers' mathematics identities have been influenced by their experiences as learners of mathematics, Tansy shows why mathematics teachers have to consider teacher development processes, including preservice courses, as major construction sites in the shaping of such identities. Her chapter highlights a prevailing discourse of deficiencies in teachers' subject knowledge that surrounds specifically mathematics in primary teacher education courses. This perception of a deficiency is brought into 'reality' via the powerful discourse that surrounds the notion and creates a situation where preservice primary teachers are 'forced' to confront the mathematics they know and how they know it, to question (arguably in a negative way) their own relationship with mathematics. In essence, there seems almost a predefined or 'designated identity' (Sfard & Prusak, 2005b) for mathematics which preservice primary teachers are expected to face up to and/or indeed 'fit' into, with implications for the relationships they then develop not only with mathematics, but also with mathematics teaching. Tansy also illustrates how dominant (political) discourses can come to determine educational policy and how **this** particular discourse can come to influence the internal psyche of teachers and potentially strait jacket the mathematical identities they grow into and take with them to classrooms.

Choice

Choice is a concern throughout this book and not just in those chapters specifically dealing with it. In particular, the rationale for this book is, in part, the large number of students who do not continue with mathematics beyond compulsory education. Identities and students' relationships with mathematics are brought to the fore in considerations of why seemingly large numbers of students 'choose' to leave mathematics. By making the choices that they do, students make statements of self-understandings of who they are. However, this is but one aspect of choice. Heather Mendick, Marie-Pierre Moreau, and Debbie Epstein (see Chapter 7) speak of 'compulsory choice' in delineating issues surrounding the proliferation of educational choices now available to students, suggesting that some may feel compelled to make the choices they do, rather than these being in any way 'free'. Their chapter highlights Bourdieu's point (cited in Wacquant, 1989) that people do not choose the principle of whatever choices they make. As well as choices being part of a larger structure, people come to choices with histories of their own which in various ways directs the decisions that they make. The chapters on choice bring out questions of whom or what is doing the choosing, of restrictions on choice, and how parameters on choice have important and arguably unique meanings when considered in the context of mathematics, where discourses of power and prestige are prevalent. In relation to learning and doing mathematics, it may well seem to some students that 'choice' is not an option open to them, thus influencing the nature of their participation in and relationship with mathematics.

Although identity as relationship comes through most strongly in the psychoanalytic chapters, identity as participation is central to those written from within a sociocultural perspective. Wenger (1998), for example, linked the concept of identity with various levels and forms of participation and nonparticipation, for example, marginal and peripheral. In the chapters where compulsory stages of mathematics are being considered, there seems to be a striking commonality in the stories presented in the limited choice available to students about the nature of their participation in learning mathematics. Further, this appears to be more the case for students whom the education system has labelled as being of 'low mathematical ability' (e.g., in the chapters by Jeremy Hodgen and Rachel Marks [Chapter 4] and by Tamara Bibby [Chapter 11]). That is, 'low-ability' students are offered restricted modes of participation in school mathematics, are presented with a more procedural diet of mathematics, and so are offered restricted ways to be in mathematics. As Candia Morgan (Chapter 9) and Birgit Pepin (Chapter 10) show, they are thus allowed marginal access to enter mathematical spaces and are confined to the periphery of such spaces. In essence, they do not belong. But what are the consequences of this for their identities? I take up this question in the next section which addresses the question of whether mathematical identities matter.

DO MATHEMATICAL IDENTITIES MATTER?

The chapters of this book have used several constructions of identity as associated with mathematics. As examples, Tamara Bibby (Chapter 11) uses the term 'learner identities in mathematics' and Stephen Lerman (Chapter 13) uses that of 'pupils' identities in relation to mathematics' and also refers to 'school mathematics identities', a construction which contains a recontextualisation of mathematics based on the effects that major actors have on what mathematics gets taught and learned in schools. All these terms present different shades of meaning which are explored in more detail by Paola Valero in the next chapter (Chapter 18). Here, in looking at the term 'mathematical identities', I want to call attention to how the qualifier **mathematical** suggests that the concern is primarily with the self-understandings that people form of and/or for themselves that might be considered mathematical. Further, with this construction, there is an assumption that all who have experienced learning mathematics in school have one of these, a mathematical sense of self-identity, and have in some way formed a relationship or connection with mathematics. Again, the nature of participation in mathematics impinges on the relationship/identity a person develops with the subject. However, between the poles of participation and nonparticipation, Holland et al. (1998) suggest that there may be cases of weak forms of participation; for mathematics, this could mean that the relationship a person forms with the subject is neither strongly positive nor negative, so they do not form much of an identity with mathematics. I examine this possibility in two ways.

First, if one accepts identities as byproducts of 'hard-won standpoints', then one could view the lived experiences some people have had with mathematics as being relatively weak so as to be essentially inconsequential on the nexus of multimembership that otherwise makes up their identities. The sense of self they have formed from classroom mathematics experiences has not resulted in a position or standpoint that may be described as 'hard-won'. The case of Caroline in Pat Drake's chapter (Chapter 14) appears to provide an example here, where the mature preservice teacher contrasts her present experience of mathematics against that of her younger student experience, describing her younger self as not being interested in mathematics, which, although negative, is a relatively bland description. This position is different from and not to be confused with that of people whose relationship/identity with mathematics has been one of disillusionment, bewilderment, and so on, such as Louise told in the chapter by Mark Boylan and Hilary Povey (Chapter 5). In this latter case, the identity is negative, but not weak.

Second, there may be a breakdown occurring within students' relationships with mathematics, whereby the way of learning they find on offer in school mathematics classrooms does not fit within the nexus of identities in the making they bring with them, so that an identity that may be described as **mathematical** is never formed. Thus, in order to 'be' in mathematics, some students find that they have to deny other aspects of their identities, and some students are not willing to or do not make that sacrifice. The learning of mathematics will suffer in such circumstances, and the student then never forms (much of) a mathematical identity.

That said, mathematics, perhaps more than other subjects in the school curriculum, appears to have a capacity for generating emotional/psychological extremes both across people and within the same person. The relationships that people form with mathematics do tend to be strongly positive and/or strongly negative. This perception of mathematical identities being usually either strongly positive or strongly negative, and rarely weak, appears to be in keeping with the binary discourse associated with mathematics explored in Jenny Shaw's chapter (see Chapter 8)—that mathematics is one thing or the other, with little room for in-betweens. Further evidence of the emotional extremes that mathematics generates is given in Hoyles' (1982) study, in which 14-year-old students gave accounts of what constituted for them good and bad learning experiences. They gave more mathematics-related examples of both good and bad experiences than for any other subject area, although the study was not mathematics-based. In this volume, the autobiographical accounts provided by Laura Black, Heather Mendick, Melissa Rodd, and Yvette Solomon with Margaret Brown (see Chapter 3) and by Pat Drake (Chapter 14) provide support for the notion that mathematics generates binary emotional states within the same person.

In concluding this section, I note that, although a myriad of things can mediate what mathematics is taught and learned in school, Laura, Heather, Melissa, and Yvette with Margaret posit the idea that mathematics itself may act as a mediator in a person's other relationships, including that with themselves. This is an alternate perspective not widely acknowledged in the literature. From this perspective, one may see the human agent as having to 'answer' a **material** agency of the discipline of mathematics (Pickering, 1995) in day-to-day interactions, separate from answering the disciplinary agency within mathematics itself. The eventuality of this for some individuals is illustrated by the narratives provided in the chapter for Rachel, Nikki, and Zoë. Despite their apparent success in mathematics, these women all tell stories of feeling defenceless in the face of mathematics, and in some ways having to find a space within their nexus of identities for this 'new' position, having to find mechanisms via which to 'answer' this material agency of mathematics, for example, via teaching which allowed 'other ways of being with maths' for Zoë. It is as if mathematics strips people bare eventually (for many people sooner, for others later), bringing them to a place where they are 'forced' to confront, question, and even reinvent their relationship with the subject. In her chapter, Jenny Shaw (Chapter 8) notes that many people who go further into mathematics eventually reach a point of feeling themselves failures. The strength of mathematics in playing this role over and over again brought me to question: 'Why mathematics?' The next section addresses this question.

WHY MATHEMATICS?

If one accepts the social character of identity, it may seem somewhat paradoxical to consider the concept of identity as associated with **mathematics**, given the relatively widely held view that mathematics is not 'social'. In reading this book, I was repeatedly struck by the question, 'Why mathematics?' Another question which expands on this is, 'Why is it or why has it become important to consider the idea of identity in relation to mathematics, its teaching and learning?' I deal with these two questions in turn.

With regard to the question 'Why mathematics?', a combination of curriculum and social issues is at play. Mathematics occupies a prestigious position in the school curriculum of most developed and developing countries, often twinned with the home language(s) of the country in government policy. Within and beyond school, Anna Sfard notes in her chapter the increasing use of numbers to characterise persons, in particular to succinctly describe their educational achievements. A similar use of numbers is illustrated by Jeremy Hodgen and Rachel Marks (see Chapter 4): Primary students come to describe themselves and their classmates in terms of the number levels of the English primary national curriculum. Thus, numbers and, by extension, mathematics are being used more in everyday life as signals of a person's identity.

Furthermore, mathematics is increasingly being seen as essential for a nation's technological advancement and has been associated with an individual's economic circumstances postschooling. Thus, mathematics knowledge and qualifications are valued both nationally and individually. Yet such qualifications continue to elude a marked proportion of people (perhaps more so in Western countries), and so mathematics has been used as a critical filter (see Schoenfield, 2002) in many of these countries. Gates and Vistro-Yu (2003) have noted that societies have imbued mathematics with significant symbolic and cultural capital, which then has implications for its position and function in the school curriculum. Specifically, these authors have stated that:

> *school mathematics* plays a significant role in organising the segregation of our society [. . .] keeping the powerless in their place and the strong in positions of power [. . .] an accusation that you 'can't do maths' [. . .] is a positioning strategy [. . .] It locates you as unsuccessful and lacking in intellectual capability [. . . and] on the edge of the employment and labour market [. . .] Mathematics education thus serves as a 'badge of eligibility for the privileges of society'. (Gates & Vistro-Yu, 2003, p. 49, italics original)

A major point here is that these functions are potentially the byproducts of **school mathematics**. This quote too points to far-reaching effects of the identity work of mathematics in schools with implications for the identities people may find (immediately) available to them beyond school. Accordingly, it is seemingly the potent mix of mathematics with **school** that has ascribed on mathematics the identity work (to do with power and powerlessness, positioning, intellectuality, employability, etc.) described by Gates and Vistro-Yu, in that schools are an enabling means for mathematics to work or be worked in such ways. The quote arguably confers on school mathematics an agency or life of its own, but one ought to keep in mind that this is a value that society has given to mathematics so that it is essentially a 'cultural arbitrary' (Bourdieu & Wacquant, 1992, p. 170), albeit one that is deeply embedded in our social structures. It also raises a question of how mathematics might be different if it had a different relationship with schooling, a question also suggested by Peter Winbourne (Chapter 6).

But why would mathematics serve as the locus for such values and responsibilities? Bourdieu (1998) has provided a perspective that suggests a function of mathematics within the reproductive role of schools, stating that:

> with a psychological brutality that nothing can attenuate, the school institution lays down its final judgements [. . .] from which there is no appeal, ranking all students in a unique hierarchy of forms of excellence, nowadays dominated by a single discipline, mathematics. (p. 28)

This quote and that from Gates and Vistro-Yu may seem a particularly strong and harsh indictment of school mathematics. But, that this is the experience of some students and teachers is clearly illustrated across several chapters in this volume and in notions of being 'done to' by mathematics. For example, Jenny Shaw (Chapter 8) describes some students' experience of mathematics as that of 'a wilfully misunderstanding object', once again seeming to give mathematics a life of its own. Mark Boylan and Hilary Povey's (Chapter 5) account of Louise also provides a poignant example in her references to a mistrust of numbers and of them not working for her because they had 'a mind of their own'. It then becomes easier to see how the identities formed in situ from such experiences of school mathematics become hard-won standpoints and may then potentially serve as places of refuge that some students resort to whenever 'mathematics' is mentioned, with negative consequences for processes of learning. Although these may be identities formed in (relation to) mathematics, these identities are not what we would desire as mathematics educators with concerns for issues of social justice.

I now turn to the second question: 'Why is it or why has it become important to consider the idea of identity in relation to mathematics, its teaching and learning?' In addressing this question, I bring aspects of pedagogy and teacher development to the fore. The mathematical identities that teachers form of and for themselves have largely (but not solely; see extracts in Pat Drake's and Tansy Hardy's chapters [Chapters 14 and 16, respectively]) been formed as hard-won byproducts of their experiences as learners of mathematics. As I noted in the Introduction to this piece, identity formation is an ongoing and evolving process, and some teachers have been able to author new mathematical identities for themselves from experiences of teaching mathematics and teacher preparation programmes. This is a crucial point—that as well as learning being a site for mathematics identity development, so must also teaching and teacher development processes be seen as such sites. However, it is arguably the case that this opportunity for renegotiated understandings of self is missed for some teachers, and that a marked proportion of them bring psychic baggage from their experiences as learners of mathematics to bear on their teaching and other pedagogic practices (see e.g., Bibby, 1999). This has implications for the relationship with mathematics that students see demonstrated before them day to day in mathematics classrooms, cultural worlds which provide for students the scope of mathematical identities they see as being available to them. Inarguably, this also has implications for the mathematics learning that may occur in classrooms. Thus, it has become important to consider issues of identities in connection with mathematics teaching and learning, as many students form a relationship with mathematics mediated by the relationship they have formed with their mathematics teacher. The cases of students Sabrina and Sally with teacher Miss South in Tamara Bibby's chapter (Chapter 11) and of students Craig and Ellen with the teacher in Pauline Davis and Julian Williams' chapter (Chapter 12) provide contrasting examples. There is a need for more teacher education programmes to become aware of and appreciate the significance of this connection, especially as it concerns the relationships and ultimately the identities that students will form with mathematics, given issues featured in dealing with the question 'Why mathematics?' earlier. Indeed, Moses and Cobb (2001) contended that access to quality mathematics education is a civil rights issue as the technological shift that has occurred in recent times makes mathematical literacy crucial for economic advancement.

IN CONCLUDING

The perspectives via which mathematical identity is explored interweave (with) each other. Although each chapter is written largely from one of the three perspectives, they also draw on or allude to influences from other perspectives. Peter Winbourne (Chapter 6), for example, whose chapter is written from a sociocultural perspective, writes of the discourse of ability grouping which in the UK context is seen seemingly as **the right way** (compare Apple, 2001, and the use of the word 'right') of 'doing school'; essentially, this discourse comes to define cultural practice. Sociocultural and discursive perspectives on mathematical identities can be seen as overlapping in broader educational processes. Social and cultural practices become part of a general milieu of what happens in 'formalised' mathematics, perhaps more so in the earlier stages of education; often these social and cultural practices have their roots in powerful discourses that have come to surround (the teaching and learning of) mathematics. Beyond school, such practices also become part of the landscape of how mathematics is used in society to define people. Such sociocultural practices and discourses can bring to the fore otherwise unconscious aspects of a person, often triggering psychic and emotional responses which then come to constitute part of the 'hard-won standpoints' which lead to the self-understandings that a person forms.

Given the ongoing concerns in mathematics education outlined and the implications for equity and access (e.g., Moses & Cobb, 2001), as well as the disconnection and disillusionment a marked proportion of people feel with their school mathematics experience, one can begin to see why it has become important to address questions of identity as related to mathematics, its teaching, and learning. Mark Boylan and Hilary Povey (Chapter 5) have noted that emotional aspects have not received as much research attention as other aspects, but that they are 'fundamental' to students' experiences of and in mathematics. Other chapters in this book also illustrate how such matters become involved in choices in and beyond schooling and the life possibilities then available to people. These considerations have also become important in some governments' policies because it is felt that the exodus from mathematics will impact on a country's technological development. If one accepts the contention of this book—that there is an inextricable connection between learning and identity—then considerations of identity ought to feature in any effort to foster mathematics learning. It is important that these considerations impact what happens in schools because these are places where much mathematics identity formation takes place.

If learning is a process of identity-in-the-making, then denying students access to the multidimensionality of their identities is to deny them access to learning. This book shows that this denial occurs often in mathematics teaching, and this suggests a source of the disillusionment that many have with the subject. Pauline Davis and Julian Williams (Chapter 12) illustrate that mathematics teaching and learning can be structured in ways that allow for a fit within the nexus of multimembership of identities that students bring with them. Further, as highlighted by Barbara Jaworski (Chapter 14), teaching is neither a necessary nor a sufficient condition for learning to occur. What ought to be more prominent in teacher preparation and development programmes, and borne in mind by all who would enter a classroom to teach, is that the primary aim of education ought to be for students to learn, and not for teachers to teach. Paying more careful consideration to issues of identity in mathematics education would seem to be a way to keep learning at the forefront of what happens in classrooms.

18 Participating in Identities and Relationships in Mathematics Education

Paola Valero

Many hours of concentration, reading, and thinking around *Mathematical Relationships in Education: Identities and Participation* were behind the formulation of some simple questions: What could be significant for me to write as a reaction to the many voices, perspectives, assertions, and contentions in the 15 core chapters in this book? What do I think I 'learned' from the book? How did this book touch or change me? How can I be a participant in the discussions that are behind this collection and the issues that will arise every time a new reader opens its pages?

Stop. What am I reacting to? A set of abstract formulations which may have generated one or another intellectual rejoinder? While reading, it helped me to imagine each one of the authors talking with me, as many in fact have done at one or another time, establishing a relationship with me around the contentions in their chapters. It seemed quite contradictory to the strong messages sent in this book to engage this exercise in a merely rational way, without trying to play with and feel in my body the ideas (and discourses) that the authors are bringing forward: Learning (mathematics) is a matter of transformation of our identities through the relationships that we establish with/in our social world, and not only a matter of conceptual change and development. The process is full of contradictions, continuities and discontinuities, feelings, and mundane matters of human flesh and blood that are normally excluded from discourses dealing with mathematics (education). I am writing a reaction to the private meeting between the authors and me, as a close-to-40-year-old woman, mother, wife, Colombian and almost Danish, academic, political scientist, but now mathematics education researcher, engaging them in how their words moved me in different ways. Now it feels that I am participating in exploring identities and relationships in mathematics education.

IDENTITY AND RELATIONSHIPS: WHAT CAN BE NOTICED?

Laura, Heather, and Yvette in their role as editors point to the two central notions in the book: identity and relationships. I am also interested in these two notions. Even if in my research the term 'identity' does not feature explicitly as a concept that I make part of my toolbox, it is definitely present as an important element to be further developed (see Alrø, Skovsmose, & Valero, 2009). The term 'relationships' is more than central in my work because of the sociopolitical approach to my research in mathematics education, which conceives education and learning to be constituted in the meeting of and exchange among human beings, their practices, and institutions. More precisely, relationships among people in situations where (school) mathematics is brought into play are certainly relationships of power (see e.g., Valero, 2007).

If I were to write about what I learned from the book about identity and relationships, how could I enter into the writing? I decided to go through the book and mark all sentences where the terms 'identity/identities' and 'relationships' feature, in an attempt to make a very loose 'discourse analysis'[1] of the messages that, throughout its chapters, the book as a whole was conveying about these two notions. Here follow some of the observations resulting from this exercise.

A first remark is that the term 'identity' featured most in chapters within the sociocultural approach, less in chapters within the discursive approach, and rarely in chapters within the psychoanalytic approach, whereas the term 'relationships' featured strongly in those chapters adopting a psychoanalytic approach and rarely in all the others. There are few chapters that explicitly link the two notions, and from that it follows that there are very few sentences where both terms appear connected. Is this a coincidence or does the different featuring of the terms mean something about the three theoretical perspectives represented in the book? Which are the commonalities and differences among the sociocultural, the discursive, and the psychoanalytic approaches? Most important, what makes them, collectively, relevant to addressing the overall question formulated by Laura, Heather, and Yvette in their preface: 'What can a focus on identities and relationships bring to understanding issues of inclusion in and exclusion from mathematics?'

Let me concentrate on the notion of identity. Which ideas seem to be communicated to the reader? First of all, the chapters talk about some particular identities. Whose identities? Those of learners or students (of mathematics) and those of (mathematics) teachers in the classrooms the authors have seen, been part of, or imagined. Very seldom do chapters reveal much about the (mathematics) education researchers' identities or, in other words, the identities of the people who wrote these texts. Pat's chapter (see Chapter 14) is an exception because she uses her own mathematical story to explain how she became interested in relationships as an important element in teacher education. Laura, Heather, Melissa, and Yvette with Margaret (Chapter 3) also 'present' themselves, although in an objectified way, by making their own stories into research material which is presented in anonymous form. I suppose that if I knew them well, I would be able to match Nikki, Rachel, and Zoë's pains, pleasures, and powers with the authors'. Of course, one could always look at the list of contributors in the book and read who is 'who'. However, that would not necessarily bring me as a reader closer to understanding a central element in the politics of academic knowledge (Popkewitz, 2004): Who is constructing objectified versions and interpretations of the world of others, even while using theoretical frameworks that call for the closeness between people's human, social, and affective experiences and their possibilities of seeing, reading about, and thinking about the world? I do not intend to suggest that all researchers should make their identities and relationships explicit and open to the scrutiny of the public, but rather to point out that I would have found it very informative to see the humans behind the assertions in the book in order to go beyond my thinking about how identities and relationships shape the knowing processes of the researchers.

One other little comment about this issue: I could not resist the power of **numberese** (see Anna's chapter [Chapter 2]) and found myself doing some counting. The proportion of women and men authors in the book called my attention: 5 out of 25 authors are men, and only 6 are not British. I wondered about the age of authors, but had few resources to do a good count. The second count could be explained by the fact that the book is the result of a British seminar series. But regarding the first count, is this concern for the issues and theoretical approaches presented in the book especially attractive to (predominantly British) women? I turned to *Opening the Research Text: Critical Insights and In(ter)ventions into Mathematics Education (Nolan & De Freitas, 2008), a book with a similar approach: The* proportion of male authors was 13 out of 27. In a book addressing a more general mainstream mathematics education audience, *New Mathematics Education Research and Practice* (Maasz & Schloeglmann, 2006), the proportion of men was 17 out of 22. Of course, these examples do not constitute evidence on gender, theoretical preferences, and research gazes. However, the question is whether viewing mathematics education from these perspectives brings some sets of values that resonate strongly with the experiences of women (and some men) and that have the potential for opening understandings that have not been a significant part of more (male-) dominated approaches to reading mathematics education. For me they do, and they provide a challenge to mainstream rationalisations of mathematics education.

I also felt attracted by the use of the word 'identity' in the book. I looked at the grammatical constructions around the word in an attempt to clarify the message that the chapters as a whole seem to be conveying. Which nouns are used? Or, how is identity named? There were five noun uses in the book:

1. Identity
2. Identities

3. Identity/position
4. Identity-in-practice
5. Identity work

Most of the time the term used is 'identity', simply. But some authors prefer the plural form. The third use makes identity into a synonym of position. The fourth binds identity to practice in such a way that the two words become one word. The fifth is a compound noun that unites identity to an active productive process.

Which adjectives are used? Or, how is identity characterised? The list of attributes being given to identity through the use of adjectives was much longer:

1. Grounded positional identity
2. Figured or imagined identity
3. Fixed fictionalised identity
4. Participatory identity
5. Lived identity
6. Powerful identities
7. Discourse identities
8. Social identities
9. Mathematical identities
10. School mathematical identities
11. Fragile mathematical identities

Each one of these expressions, formed by an adjective (or series of modifiers) and the term 'identity(ies)', points to a more precise association of meanings with the term 'identity'. Identity(ies) is not just simply identity(ies), but is characterised in relation to positions; to imaginations and fictions; to participation; to discourse; to life; to interaction, social context, or society; and to mathematics and school mathematics. It seems as if characterising identity demands the exercise of seeing a relationship of some kind.

Which verbal expressions are used? Or, which kinds of actions or states is identity involved in? I tried to group those verbs, and the following four categories resulted:

1. Conferred, ascribed, acquired, available, taken
2. Formed, (re)constructed, developed, constituted, emergent, lived
3. Established, brought, manifested, trapped
4. Shaped, managed, negotiated, mediated

Expressions using the first set of verbs could convey the idea that identity 'exists' outside of individuals, as if it were an object that people can use, such as a jacket that one can take on, and that other people can also put on you. The second group, in contrast, points to a meaning of identity being a process that people engage in, where the centre of the activity is the acting person and their experience. The third group collects verbs that signal a kind of rigidity and stability in identity. The fourth group insinuates that identity is a more malleable entity.

I also selected sentences proposing definitions of the notion of identity, clustered sentences that seemed to me to have commonalities, and partially reformulated the sentences. There were six different definitions pointing in slightly or decidedly different directions:

1. Identity is a sense of self.
2. Identity is about being recognised or passing as a certain kind of person in a given context, as well as about (re)constructing oneself that way.
3. Identity is a nexus of multiple forms of membership through a process of reconciliation across boundaries of practice. It is an ongoing, dynamic process through which individuals negotiate the meanings of their experience of membership to communities of practice. Such experience may be more or less localised in particular communities.
4. Identity is the result of the subject's interpellation into discourse, systems of knowledge, and practice, which construct objects. The process of subjectivity is inseparable from relations of power, as it is immersed in regulatory systems. Identity as subjectivity is the construction of a positioning of onself in discourse and of being positioned by discourse, both subject and subjected.
5. Identities are stories about who we are, with whom we belong, and what position we occupy among those who constitute our human environment. Changes in discourses bring changes in the world and also in our identities.
6. School mathematics identities are the identities that pupils are likely to develop as learners of mathematics within the recontextualisation that takes place when government, curriculum designers, textbook writers, and teachers construct the activities of the mathematics classroom.

In relation to the previous selections of nouns, adjectives, and verbs, the definitions make more visible the theoretical perspectives from which authors are viewing identity. Because the chapters in this book adopt three main theoretical perspectives—socio-cultural, discursive, and psychoanalytic the definitions are clearly aligned within at least two of the theoretical perspectives. Thus, the third definition could be labelled as 'sociocultural' and locates identity within the functioning of communities of practice as proposed by Wenger (1998). The fourth and fifth definitions are discursive, but of a different type. The fourth supposes a connection between individual identity and the institutionalised frames within which discourse operates, and it sees power as a constitutive element in the relationship between the individual and the social through the process of identity formation or subjectification. In contrast, the fifth definition of identities as the stories that we tell about ourselves is a definition within a discursive frame; the relationship between individual and the social do not seem to be seen within the frames of institutionalised power structures. The first definition of identity as the sense of self seems to emphasise the individual and his/her perception of his/her own being, which, in contrast with the other definitions, seems to be focussed on the individual and missing the social (discursive or institutionalised) interactions which seem to be an important part of the notion. The second definition, phrased following Gee (1999), seems to emphasise the situatedness of identity in social interactions.

The sixth definition is not a definition of the general concept of identity, but of a specified notion—that of school mathematics identities. It directs attention to particular beings (pupils in schools) who develop identities as mathematics learners within the framework of school mathematical activities in a classroom. These activities are here seen as recontextualisations into the classroom of the intentions and actions of different social actors (policymakers, curriculum designers, textbook writers, and teachers).

Finally, I tried to find sentences that explicitly addressed the issue of: Why is the notion useful? Or, which kind of insight does identity allow us to generate with respect to mathematics education? The following three points emerged:

1. Identity has a pivotal role in understanding the relationship between human agency and social structure.
2. It is a unit of analysis which allows us to concentrate on the whole person and their becoming, rather than part of the person and their knowing.
3. It can help us discuss the identities we might want students to acquire in the mathematics classroom.

These three points highlight important justifications for the usefulness of the concept in researching mathematics education. To the old sociological dispute on whether the key to studying the 'social' is individual actions or collectively produced structures (see e.g., Rocher, 1978), the concept of identity offers a concrete proposal: focussing on identity allows us to bridge the gap between what has been taken as individually centred social processes and the socially constructed institutionalised world. In mathematics education, a similar dispute has been going on between defenders of theories of learning focussing on the individual and defenders of learning theories privileging the social. As seen in most of the definitions of identity presented earlier, the relational nature of the concept between the individual and the social evidences that it can in fact represent a way of understanding human activity, in particular human learning, as a constant relationship between the individual and the social world. The social world here is understood in a variety of ways from microsocial interaction to more macrosociological structures. The second assertion about identity opening the possibility of seeing human beings as whole beings in a process of becoming, rather than as cognitive agents, signals the usefulness of researching the complexity of mathematics education as processes that involve much more than what research has traditionally taken as a focus of study. The third assertion points to a normative dimension: that of what we would like to happen with the identities of people in relation to mathematics in classrooms. This means that identity can also be used in devising intervention techniques or pedagogical development aimed at bettering the possibilities of mathematics learning.

As for the notion of relationship(s), the exercise of looking in the text for the sentences where the word was used as a concept did not give as many results as for the word 'identity'. This probably has to do with the fact that, as mentioned previously, the notion is more central in chapters adopting a psychoanalytic approach, and those represent one third of the chapters in the book. The exercise of finding some kind of categorisation of the sentences and terms encountered was much more difficult for me, and therefore the following reflections seem less organised. I tried to group the sentences containing the term in four different clusters pointing to some aspects of the notion.

First, when indicating a concept, the noun 'relationship' was both in singular and plural, but occurred mostly in plural. This use may indicate that relationships are repeated linkages and not just one coincidental rapport. Relationships point to patterns of connections and associations, and therefore, to certain stability in encounters with others or with the material world that are influential for individuals' lives and possibilities for knowing and learning. The term appears indicating connections among people—a child and the mother, the pupil and the teacher, one person and 'the other', one person and his/her significant others, and so on—or indicating connections between people and objects—one's relationship with the breast, one's relationship with the school, one's relationship with mathematics, and so on. The term was also found together with different adjectives that indicated the type of relationships: There could be close, conflictive, emotional, intimate relationships among people or between people and mathematics.

Second, there was a group of sentences providing clarification about the theoretical perspective behind the views on relationships:

1. Relationships are hard work; they involve knowledge and thinking that go beyond the rational.
2. There is a one-way, 'doer/done to' relationship and one characterised by two-way 'intersubjectivity'.
3. The 'object relations' theory stresses the importance of relationships and the experience of them as formative and as shaping the person emotionally. The term 'object' generally means a person or part of a person; and the core assumption is that the search for relationships and someone to have one with is the driving force of life.

4. The early experience of being understood or misunderstood can be particularly profound and affects not only relationships with people, but also with institutions, schools, and workplaces, for example, and with bodies of knowledge such as mathematics.

The first and second sentences provide more characterisation to the concept. Defining relationships as hard work suggests that people invest effort into relationships and that they are not simply the result of a fortuitous meeting or of a rational interaction among people. They appeal to other dimensions of human beings, such as emotions. Furthermore, there are different types of relationships, unidirectional and bidirectional, where the effect on the people involved is not reciprocal and building toward intersubjectivity. The third sentence presents a particular theory on relationships and its significance for human life. The fourth sentence adds to the previous one by pointing to the experience of being understood and its effects on how people establish relationships with other people, institutions, and bodies of knowledge. Parents—and close significant others—are also mentioned as key people whom one establishes relationships with, and the impact of these primary relationships with the mental—and also emotional—development of the individual.

Third, a group of sentences refer to the significance of relationships for learning and pedagogy:

1. Children know that what and how they know is intimately bound up in relationships.
2. Relationships and subjectivities frame and constitute what is known and knowable.
3. We have to develop relationships to learn and to enable others to learn.
4. Pedagogy is about building relationships.

These statements stress the idea that, in a learning process, possibilities for engaging with the content of learning cannot be seen outside of the human exchanges enabling it. But such human exchanges are not simply 'interactions of a social nature', but are deep emotional relationships amongst the people involved in the learning. Although most pedagogical work may still focus on how to bridge the gap between the knowledge of adults and experts and the knowledge of learners, these sentences suggest that the role of pedagogy is to facilitate relationships amongst people from which learning can be enabled.

Finally, a fourth cluster gathered sentences addressing students' and teachers' relationships with mathematics:

1. We should take account of people's emotional relationship with mathematics knowledge, as well as their intellectual grasp of it. However, initiatives to develop mathematics teachers are frequently couched in terms of mathematics subject knowledge.

2. People's relationships with mathematics are very likely to lead to different approaches to teaching. Teachers renegotiate their relationship with learning and so teaching mathematics.
3. This is a switch of focus that foregrounds not the individual and their 'choices' and 'abilities', but the ways that people are assembled as (not) choosing mathematics and as un/able through patterns of relationships, materiality, and so on.
4. The different relationships to mathematics available to students mirror the systemic assumptions about the possible trajectories of the target groups. A vicious circle is thus constructed, in which the student who is judged to be less successful at mathematics is provided with resources that distance them even more from powerful mathematical practices.

These four groups of sentences point in different (but still related) directions. The first argues for the importance of considering people's emotional relation with mathematics side by side with their intellectual engagement with it and pleads for considering this dimension seriously when educating mathematics teachers. The second sentence relates people's relationships with mathematics and people's identities through emotion. Such emotional association is also at the basis of people's approaches to teaching. Teachers face the task of revising and reconstructing their emotional relationship with learning and teaching mathematics. The third sentence indicates that people's choices and abilities in mathematics are not an intrinsic characteristic of individuals, but rather the result of relationships and conditions which make possible and available for people the options of not/choosing mathematics and being good/bad at mathematics. The fourth sentence signals that there is a connection between the types of relationships that students can construct with mathematics and the types of activities and practices made available to them through teaching. People's relationships with mathematics, echoing the previous sentence, emerge in contexts of practices and conditions and are not initiated individually.

I now go back to my initial questions: What did I learn from the book? How did the book touch me? In the previous paragraphs, I have tried to present reflections connected to the exercise of taking the phrasings on identity and relationships outside of the particular contexts of the chapters. My general overview at this time seems to be that the book talks much more about identity than about relationships, that identity is a much more nuanced concept in how it is used and defined in relation to how the concept of relationships appears in the text, and that there are clear connections between ways of talking about identity and relationships and the theoretical perspectives that the book brings forward. But probably the overall message that I certainly connect with and feel strongly about is that mathematics education practices are complex social phenomena, and that generating richer understandings of their complexity demands going beyond frameworks focussing on mathematical learning to seeing whole human beings engaging in the construction of their subjectivities through relationships with 'mathematics'.

As I come to this point, I realise that the chapters probably did not challenge the concept of mathematics. A new exercise of 'discourse analysis' could probably help clarify whether the focus on identity and relationships also has implications for how authors conceptualise mathematics. Is it in fact a concept that is examined in the book, or is it treated as a 'black box'? Should the theoretical perspectives adopted in the chapters bring new light into what is understood by mathematics?

Another strong question—one among very many—still remains, and it is whether I felt that the book, with a focus on identities and relationships, allows me to construct a better understanding of issues of in(ex)clusion in mathematics education. Let me take this question as an entry point into a final reflection.

IN(EX)CLUSION WITH IDENTITIES AND RELATIONSHIPS

A strong part of my research agenda has been the concern for the political dimensions of mathematics education—more concretely, how and why mathematics education is seen as being 'powerful' and what such views imply for the participation of different students and teachers in the social-political practices of mathematics education. The issue of inclusion and exclusion is, therefore, dear to me. As I read and engaged in (virtual, imagined) conversations with the authors, I asked each one about whether their readings of practices and experiences in schools, colleges of education, and society in general helped me come to a better grasp of power in mathematics education. Many authors replied positively and showed me the mechanisms through which multiple practices were constituting individual and collective ways of acting and talking that had the potential to position individuals and communities in in(ex)cluded standpoints. What did these authors tell me?

I start by expounding briefly the construct of the 'network of mathematics education practices' (Valero, 2007), to which I am relating the authors' claims. For me, mathematics education can be defined as a series of social practices carried out by different people in different sites, where the meaning of the teaching and learning of mathematics is constituted in particular historical conditions. Those social practices happen not only in the classroom, where teachers and students interact around mathematical content, but also take place in, for example, parents' demands on schools, local community educational needs, international or national educational policymaking in mathematics, teacher education, textbook production, the labour market, mathematics education research, mathematics research, youth culture, public views and discourses of mathematics, and so on. These various sites of practice are sometimes loosely and sometimes tightly interconnected depending on particular historical circumstances, but they are all implicated in the construction of the multiple meanings ascribed to the teaching and learning of mathematics in a given time and location. Defining mathematics education in terms of the network of mathematics education practices points to the sociological and political complexity of mathematics education. Thus, this view has implications for how I decide to engage in the task of researching mathematics education because it allows me to see classrooms—the traditional site of research in mathematics education—in relation to many other sites of practice in an attempt at uncovering the microphysics of power in everyday social interactions and practices (Christensen, Stentoft, & Valero, 2008a, 2008b). Such a move in research is, from my point of view, a way of tackling seriously the need for better understandings of how mathematics education is implicated in in(ex)clusion processes in society.

Many chapters in the book illuminate different sectors of the network of mathematics education practices. Jeremy and Rachel (Chapter 4) connect students' and teachers' construction of identity with the notions of mathematical ability institutionalised through the practice of ability grouping and the practices of assessment and examination in the English educational system. They are able to describe the mechanisms through which certain notions of mathematical ability are created and how they impact students' and teachers' perceptions of themselves as good/deficient learners. This chapter on practices is complemented by the chapter by Laura, Heather, Melissa, and Yvette with Margaret (Chapter 3) about their own experiences—on how subjects invest in discourses that protect an exposed and anxious self against the brutal attack of classification and selection systems embedded in the way teaching and learning practices of mathematics are configured.

Pauline and Julian (Chapter 12) connect different sites of practice in the network of mathematics education: classroom practices of learning with youth and out-of-school culture. They show how practices allowing a hybridity in ways of talking about mathematics open up meaningful participation and identity construction for students in mathematics, thanks to the possibility of linking to their world. 'Bringing reality into the classroom', from this point of view, is not about using school problems with a referent to some distant kind of students' reality, but rather a serious recognition of the fact that as human beings we have multiple memberships in a variety of communities and that they are all part of our participation in the world. Therefore, bringing real peer talk into the mathematics classroom is a powerful way of breaking with the exclusion that a pure maths talk would normally bring.

Tamara (Chapter 11) makes a very strong case for reconceptualising pedagogy focussing on the clear recognition of the significance of human relationships—and all the emotionality they bring—as a necessary condition for learning. Her proposal presents a direct challenge to the dominant idea that the only function of relationships between teachers and students is to provide a vehicle for mathematics learning as conceptual change. Tamara's point on the importance of emotional relationships being the bearer of learning resonates with Jenny's (Chapter 8) reading of people's affections and disaffections with mathematics. However, Jenny knits links between people in relation with mathematics and people's experiences in the world, meetings with others, and encounters with discourses. Jenny adds an insight that for me complements Tamara's strong case for a different pedagogy: Learning is not experienced as linear, gradual, and incremental—as dominant theories of learning invite us to think—but as 'jerky, episodic, and beset with loss as much as gain'. This view of learning and relationships with mathematics is also exemplified by the plurivocality, disruptions, and discontinuities in feelings with mathematics that Mark and Hilary (Chapter 5) grasp in the narrative of a female student, Louise. For me, Jenny's proposal, as well as Tamara's and Mark and Hilary's illustrations, really challenge central discourses about the meaning of teaching and learning mathematics in, for example, the wave of constructivist reform agendas put forward by policymakers, researchers, teacher educators, and professional organisations alike.

Birgit's contribution (Chapter 10) is another example of how different sites in the network of mathematics education practices set the frame for what is available for different students to be and identify with. By examining textbooks—and thereby bringing into the scene textbooks writers' agendas for mathematics education—and their introduction into classroom routines, Birgit shows that mathematics classroom practices are sites open to the influence of other engineering forces that, indirectly, have significance for the possibilities of learning and teaching. But, of course, there are many other shaping forces in mathematics education. Candia makes a brilliant analysis of the way in which official regulatory discourses, which are part of mathematics education systems, impact 'directly and powerfully on individual and collective experience within the institutional structure of the school'. Taking the curricular texts in mathematics in England at this time in history, she evidences how they produce discourses that frame possibilities for practice for both students and teachers and, thereby, set the basis for the instauration of clear mechanisms of in(ex)clusion.

Peter (Chapter 6) exemplifies the workings of the network of mathematics education practices within the school organisation of mathematics education. All different participants in a school community—students, teachers, school administrators, parents, and local authorities—are collectively responsible for the 'choices', decisions, and transformations that set the framework of practice in schools. Although Peter seems to apologise for the way that his broader focus on the contextualisation of children's, parents', and teachers' activities sacrifices focus on the learning and teaching of mathematics, I would say that it is precisely because of his focus on the 'context' that he is able to unpack the complexity of the functioning of mathematics education—and the meanings it has—in a school community. Peter's analysis, for me, really shows the complexity that is avoided in much of the research on mathematics school change.

Heather, Marie-Pierre, and Debbie (Chapter 7) analyse people's choice of mathematics in the framework of a neoliberal context, where it is possible for learners to 'construct their identities in relation to popular cultural representations of mathematics and mathematicians' and, I would add, the public discourses about mathematics that emerge from people's experience of mathematics in educational institutions. This chapter again connects students' choices with ideas that circulate in the network of mathematics education practices about ability, intelligence, individuality, difference, and specialness. Their choices are not only the result of those values, but they also contribute to perpetuating the labelling of students who did not choose mathematics as people who simply 'do not have what it takes' to succeed and, therefore, need to work hard. Tansy (Chapter 16) touches on another set of connected sites in the network of mathematics education practices. Mathematics teacher education is considered in relationship to policymaking, also touching on the mass media. With an examination of the multiple discourses circulating about the qualifications of teachers and how those languages entangle with the identity that teacher educators and student teachers develop, Tansy certainly brings forward a dynamic and complex analysis of the discursive field within which (mathematics) primary teacher education in England is taking place.

All these chapters helped me develop my thinking about power, in(ex)clusion, and the network of mathematics education practices. For me, the different chapters provide concrete illustrations of the interactions among people at the source of the production of numberese, the numerical discourses that are strongly characteristic of Modern, Western culture. Anna (Chapter 2) makes a fantastically clear case in identifying numberese and in pointing to its main four features of informativeness, generality, rigour, and objectivity. The existence and persistence of numberese, one of the big discourses that emerges from and is constantly (re)constructed in the functioning of the network of mathematics education practices, and its invasion of our identities, is only possible because of the human discourses, actions, feelings, and relationships that many of the chapters in this book have illustrated, evidenced, and deconstructed. The power of numberese, and of mathematics and mathematics education, is not only connected to the internal strengths and characteristics of numberese as a discourse, but also to the high valorisation that human beings, individually and collectively, have given to it in the construction of Modernity. Without numberese, it would not have been possible to establish the different processes of normalisation, classification, and governmentalisation (see e.g., Desrosières, 1998). Resisting numberese, as Anna invites us to do, implies taking more seriously power as an important element in what we do.

Finally, I really want to thank the editors and authors for having enriched me.

NOTES

1. I am not claiming to have done a 'proper' discourse analysis. I looked for similarities and patterns in the uses of the words, reaching some kind of categorization. I allowed myself to reflect on the ideas or meanings that I see behind the different clusters. I do not use direct citations, but allow myself to point to more or less paraphrased wording from the text.

References

Adler, J. (2008, March). *Empirical and theoretical reflections on researching mathematics for teaching in mathematics teacher education*. Paper presented at the Symposium on the Occasion of the 100th Anniversary of ICMI, Rome, Italy.

Adler, J., & Davis, Z. (2006). Opening another black box: Researching mathematics for teaching in mathematics teacher education. *Journal for Research in Mathematics Education, 36*(4), 270–296.

Alrø, H., Skovsmose, O., & Valero, P. (2009). Inter-viewing foregrounds. In M. César & K. Kumpulainen (Eds.), *Social interactions in multicultural settings*. Rotterdam: Sense.

Apple, M. (2001). *Educating the 'right' way: Markets, standards, God, and inequality*. New York: RoutledgeFalmer.

Archer, L., Hutchings, M., & Ross, A. (2003). *Higher education and social class: Issues of exclusion and inclusion*. London: RoutledgeFalmer.

Askew, M., Brown, M., Rhodes, V., Johnson, D. C., & Wiliam, D. (1997). *Effective teachers of numeracy (final report)*. London: King's College.

Baker, D., Bland, P., Hogan, P., Holt, B., Job, B., Verity, R., et al. (2000). *Keymaths 7^1 and 7^2*. Cheltenham: Stanley Thornes.

Bakhtin, M. (1981). *The dialogic imagination* (C. E. Merson & M. Holquist, Trans.). Austin: University of Texas Press.

Bakhtin, M. (1986). *Speech genres and other late essays* (V. W. McGee, Trans.). Austin: University of Texas Press.

Ball, D. L., & Lampert, M. (1999). Multiples of evidence, time and perspective: Revising the study of teaching and learning. In E. L. Lagemann & L. S. Shulman (Eds.), *Issues in education research* (pp. 371–398). San Francisco: Jossey-Bass.

Ball, S. J. (1994). *Education reform: A critical and post-structural approach*. Buckingham, UK: Open University Press.

Ball, S. J. (2001). Performativities and fabrications in the education economy. In C. Husbands & D. Gleeson (Eds.), *The performing school: Managing, teaching and learning in a performance culture* (pp. 210–226). London: Falmer.

Ball, S. J. (2003). The teacher's soul and the terrors of performativity. *Journal of Education Policy, 18*(2), 215–228.

Ball, S. J. (2007). *Education plc: Understanding private sector participation in public sector education*. Abingdon: Routledge.

Ball, S. J. (2008). *The education debate*. Bristol: The Policy Press.

Barnes, M. (2000). Effects of dominant and subordinate masculinities on interactions in a collaborative learning classroom. In J. Boaler (Ed.), *Multiple perspectives on mathematics teaching and learning* (pp. 145–170). London: Ablex Publishing.

Bartholomew, H. (2000, September). *Negotiating identity in the community of the mathematics classroom.* Paper presented at the British Educational Research Association annual conference, Cardiff University, Cardiff, UK.
Bartholomew, H. (2001). *Learning environments and student roles in individualised mathematics classrooms.* Unpublished doctoral thesis, King's College, University of London, London.
Bartholomew, H. (2005, July). *Top set identities and the marginalisation of girls.* Paper presented at the Fourth International Mathematics Education and Society Conference, Griffith University, Australia.
Bauman, Z. (2001). *The individualized society.* Cambridge: Polity Press.
Beck, U., Giddens, A., & Lash, S. (1994). *Reflexive modernization: Politics, tradition and aesthetics in the modern social order.* Cambridge: Polity Press.
Ben-Yehuda, M., Lavy, I., Linchevski, L., & Sfard, A. (2005). Doing wrong with words: What bars students' access to arithmetical discourses. *Journal for Research in Mathematics Education, 36*(3), 176–247.
Benjamin, J. (1998). *Shadow of the other: Intersubjectivity and gender in psychoanalysis.* New York/London: Routledge.
Benjamin, J. (2004). Beyond doer and done to: An intersubjective view of thirdness. *Psychoanalytic Quarterly, LXXIII,* 5–46.
Benwell, B., & Stokoe, E. (2006). *Discourse and identity.* Edinburgh: Edinburgh University Press.
Bernstein, B. (1971). *Class, codes and control.* London: Routledge & Kegan Paul.
Bernstein, B. (1990). *Class, codes and control volume IV: The structuring of pedagogic discourse.* London: Routledge.
Bernstein, B. (1999a). Official knowledge and pedagogic identities. In F. Christie (Ed.), *Pedagogy and the shaping of consciousness: Linguistic and social processes* (pp. 246–261). London: Continuum.
Bernstein, B. (1999b). Vertical and horizontal discourse: An essay. *British Journal of Sociology of Education, 20*(2), 157–173.
Bernstein, B. (2000). *Pedagogy, symbolic control and identity* (rev. ed.). Lanham, MD: Rowman & Littlefield.
Bernstein, B. (2004). Social class and pedagogic practice. In S. J. Ball (Ed.), *The RoutledgeFalmer Reader in Sociology of Education* (pp. 196–217). London: RoutledgeFalmer.
Bibby, T. (1999). Subject knowledge, personal history and professional change. *Teacher Development, 3*(2), 219–232.
Bibby, T. (2001). *Primary school teachers' personal and professional relationships with mathematics.* Unpublished doctoral thesis, King's College, University of London.
Bibby, T. (2002). Shame: An emotional response to doing mathematics as an adult and a teacher. *British Educational Research Journal, 28*(5), 705–722.
Bibby, T. (2006, November). *Mathematical relationships: Identities and participation: A psychoanalytic perspective?* Paper presented at the ESRC seminar, Mathematical Relationships: Identities & Participation, University of Manchester, Manchester, UK.
Bion, W. (1970). *Attention and interpretation: A scientific approach to insight in psychoanalysis and groups.* London: Tavistock Publications.
Bion, W. (2004/1961). *Experiences in groups and other papers.* Hove, East Sussex: Brunner-Routledge.
Black, L. (2004a). Differential participation in whole-class discussions and the construction of marginalised identities. *Journal of Educational Enquiry, 5*(1), 34–54.
Black, L. (2004b). Teacher–pupil talk in whole-class discussions and processes of social positioning within the primary school classroom. *Language and Education, 18*(5), 347–360.

Black, P., Harrison, C., Hodgen, J., Marshall, B., & Serret, N. (2007, September). *Riding the interface: An exploration of the issues that beset teachers as they strive for assessment systems*. Paper presented at the British Educational Research Association annual conference, Institute of Education, University of London, London, UK.

Black, P., Harrison, C., Lee, C., Marshall, B., & Wiliam, D. (2002). *Working inside the black box*. London: King's College.

Black, P., & Wiliam, D. (1998a). Assessment and classroom learning. *Assessment in Education, 5*, 7–73.

Black, P., & Wiliam, D. (1998b). *Inside the black box: Raising standards through classroom assessment*. London: King's College.

Blatchford, P., Hallam, S., Ireson, J., Kutnick, P., & Creech, A. (2008). Classes, groups and transitions: Structures for teaching and learning (Report no. 9/2). *Primary Review Research Survey*. Retrieved November 20, 2008, from http://www.primaryreview.org.uk/Publications/Interimreports.html.

Bloomer, M., & Hodkinson, P. (2000). Learning careers: Continuity and change in young people's dispositions to learning. *British Educational Research Journal, 26*(5), 583–597.

Boaler, J. (1997). *Experiencing school mathematics: Teaching styles, sex and setting*. Buckingham: Open University Press.

Boaler, J. (2000a). Mathematics from another world: Traditional communities and the alienation of learners. *Journal of Mathematical Behaviour, 18*(4), 379–397.

Boaler, J. (2000b, July). *So girls don't really understand mathematics? Dangerous dichotomies in gender research*. Paper presented at the Ninth International Congress of Mathematics Education Conference, Tokyo, Japan.

Boaler, J. (2002a). The development of disciplinary relationships: Knowledge practice and identity. *For the Learning of Mathematics, 22*(1), 42–47.

Boaler, J. (2002b). *Experiencing school mathematics: Traditional and reform approaches to teaching and their impact on student learning*. Mahwah, NJ: Lawrence Erlbaum Associates.

Boaler, J. (2008). Promoting 'relational equity' and high mathematics achievement through an innovative mixed-ability approach. *British Educational Research Journal, 34*(2), 167–194.

Boaler, J., & Greeno, J. G. (2000). Identity, agency, and knowing in mathematics worlds. In J. Boaler (Ed.), *Multiple perspectives on mathematics teaching and learning* (pp. 171–200). Westport, CT: Ablex Publishing.

Boaler, J., & Staples, M. (2008). Creating mathematical futures through an equitable teaching approach: The case of Railside school. *Teachers' College Record, 110*(3), 608–645.

Boaler, J., Wiliam, D., & Brown, M. (2000). Students' experiences of ability grouping-disaffection, polarisation and the construction of failure. *British Educational Research Journal, 26*(5), 631–648.

Boaler, J., Wiliam, D., & Zevenbergen, R. (2000, March). *The construction of identity in secondary mathematics education*. Paper presented at the Mathematics Education and Society Conference, Montechoro, Portugal; Centro de Investigação em Educação da Faculdade de Ciências Universidade de Lisboa.

Borich, G. (1996). *Effective teaching methods*. New York: Macmillan Press.

Bourdieu, P. (1998). *Practical reason: On the theory of action*. Cambridge: Polity Press.

Bourdieu, P., & Wacquant, L. (1992). *An invitation to reflexive sociology*. Cambridge: Polity Press.

Boylan, M. (2004). *Questioning (in) school mathematics: Life worlds and ecologies of practice*. Unpublished doctoral thesis, Sheffield Hallam University, Sheffield.

Boylan, M., Lawton, P., & Povey, H. (2001, July). *'I'd be more likely to speak in class if . . .': Some students' ideas about increasing participation in whole class interactions'*. Paper presented at the 25th Conference of the International Group for the Psychology of Mathematics Education, Utrecht, The Netherlands.

Britzman, D. (1998). *Lost subjects, contested objects: Toward a psychoanalytic theory of learning.* Albany: State University of New York Press.

Britzman, D. (2003). *After-education: Anna Freud, Melanie Klein, and psychoanalytic histories of learning.* Albany: State University of New York Press.

Brown, M., Askew, M., Hodgen, J., & Wiliam, D. (2009). Individual and cohort progression in learning numeracy ages 5–11: Results from the Leverhulme 5-year longitudinal study. In A. Dowker (Ed.), *Children's mathematical difficulties: Psychology, neuroscience and education* (pp. 103–126). Oxford: Elsevier.

Brown, M., Askew, M., Rhodes, V., Denvir, H., Ranson, E., & Wiliam, D. (2003, March). *Characterising individual and cohort progression in learning numeracy: Results for the Leverhulme 5-year longitudinal study.* Paper presented at the American Education Research Association Conference, Chicago, IL.

Brown, M., Brown, P., & Bibby, T. (2008). I would rather die': Attitudes of 16 year-olds towards their future participation in mathematics. *Research in Mathematics Education, 10*(1), 3–18.

Brown, S., & McIntyre, D. (1993). *Making sense of teaching.* Buckingham: Open University Press.

Brown, T. (1997). *Mathematics education and language: Interpreting hermeneutics and post-structuralism.* Dordrecht: Kluwer Academic Publishers.

Brown, T., Jones, L., & Bibby, T. (2004). Identifying with mathematics in initial teacher training. In M. Walshaw (Ed.), *Mathematics education with/in the postmodern* (pp. 161–179). Greenwich, CT: Information Age.

Bruner, J. (1985). Vygotsky: a historical and conceptual perspective. In J. V. Wertsch (Ed.), *Culture Communication and Cognition: Vygotskian Perspectives* (pp. 21–34). Cambridge: Cambridge University Press.

Burgess, R. G. (1984). *In the field: An introduction to field research.* London: Allan & Unwin.

Burton, L. (1995). Moving towards a feminist epistemology of mathematics. In G. Kaiser & P. Rogers (Eds.), *Equity in mathematics education: influences of feminism and culture* (pp. 209–225). London: Falmer.

Burton, L. (2004a). *Mathematicians as Enquirers: Learning about Learning Mathematics.* Dordrecht: Kluwer/Springer.

Burton, L. (2004b). 'Confidence is everything'—perspectives of teachers and students in learning mathematics. *Journal of Mathematics Teacher Education, 7*(4), 357–381.

Butler, J. (1997). *The psychic life of power.* Stanford, CA: Stanford University Press.

Butler, J. (2008). Sexual politics, torture, and secular time. *The British Journal of Sociology, 59*, 1–23.

Buxton, L. (1981). *Do you panic about maths? Coping with maths anxiety.* London: Heinemann.

Cane, A. (2007, August 25). 'Maths hysteria'. How can books about mathematics be made accessible? It helps to add tales of discoveries, duels and drunkenness to the algebra. *Financial Times Magazine,* pp. 30–31.

Canham, H. (2006). 'Where do babies come from?' What makes children want to learn? In B. Youell (Ed.), *The learning relationship: Psychoanalytic thinking in education* (pp. 7–19). London: Karnac.

Carr, M. (2001). A sociocultural approach to learning orientation in an early childhood setting. *Qualitative Studies in Education, 14*(4), 525–542.

Cestari, M. L., Daland, E., Eriksen, S., & Jaworski, B. (2006). Working in a developmental research paradigm: The role of didactician/researcher working with teachers to promote inquiry practices in developing mathematics learning and teaching. In M. Bosch (Ed.), *Proceedings of the Fourth Congress of the European Society for Research in Mathematics Education* (pp. 1348–1357). Sant Feliu de Guíxols, Spain: Universitat Ramon Llull.

Charmaz, K. (2000). Grounded theory: Objectivist and constructivist methods. In N. K. Denzin & Y. S. Lincoln (Eds.), *Handbook of qualitative research* (2nd ed., pp. 509–535). London: Sage.

Chevallard, Y. (1992). Fundamental concepts in didactics: Perspectives provided by an anthropological approach. In *Recherches en Didactique des Mathématiques: Selected Papers* (pp. 131–167). Grenoble: La Pensée Sauvage.

Chouliaraki, L., & Fairclough, N. (1999). *Discourse in late modernity: Rethinking critical discourse analysis*. Edinburgh: Edinburgh University Press.

Christensen, O. R., Stentoft, D., & Valero, P. (2008a). A landscape of power distribution. In K. Nolan & E. De Freitas (Eds.), *Opening the research text: Critical insights and in(ter)ventions into mathematics education* (pp. 147–154). New York: Springer.

Christensen, O. R., Stentoft, D., & Valero, P. (2008b). Power distribution in the network of mathematics education practices. In K. Nolan & E. De Freitas (Eds.), *Opening the research text: Critical insights and in(ter)ventions into mathematics education* (pp. 131–146). New York: Springer.

Cobb, P., & Hodge, L. (2002, July 7–12). *Learning, identity and statistical data analysis*. Paper presented at the International Conference on Teaching Statistics, Cape Town, South Africa.

Coben, D. (2000). Mathematics or common sense? Researching 'invisible' mathematics through adults' mathematics life histories. In D. Coben, J. O'Donoghue, & G. E. Fitzsimons (Eds.), *Perspectives on adults learning mathematics: Research and practice* (pp. 53–66). Dordrecht: Kluwer.

Cochran-Smith, M., & Lytle, S. L. (1999). Relationships of knowledge and practice: Teacher learning in communities. *Review of Research in Education*, 24, 249–305.

Cooper, B. (1994). Secondary mathematics education in England: Recent changes and their historical context. In M. Selinger (Ed.), *Teaching mathematics* (pp. 5–26). London: Routledge.

Cooper, B., & Dunne, M. (2000). *Assessing children's mathematical knowledge: Social class, sex and problem-solving*. Buckingham: Open University Press.

Cooper, D. (2006). *Talk about assessment: Strategies and tools to improve learning*. Toronto, ON: Thomson Nelson, Government of British Columbia.

Cronin, A. (2000). *Advertising and consumer citizenship: Gender, images and rights*. London: Routledge.

Daland, E. (2007). School teams in mathematics, what are they good for? In B. Jaworski, A. B. Fuglestad, R. Bjuland, T. Breiteig, S. Goodchild, & B. Grevholm (Eds.), *Learning communities in mathematics* (pp. 161–174). Bergen: Caspar.

Damarin, S. (2000). The mathematically able as a marked category. *Gender and Education*, 12(1), 69–85.

Davies, B. (2003). Death to Critique and Dissent? The Policies and Practices of New Managerialism and of 'Evidence-based Practice'. *Gender and Education*, 15(1), 91–103.

Davis, P. (2007a). Discourses about reading among seven and eight year old children in classroom pedagogic cultures. *Journal of Early Childhood Literacy*, 7, 219–252.

Davis, P. (2007b). How cultural models about reading mediate classroom (pedagogic) practice. *International Journal of Educational Research*, 46 (1), 31–42.

DCSF. (2006). Social mobility: Narrowing social class educational attainment gaps: Supporting materials to a speech by the Rt. Hon. Ruth Kelly MP, Secre-

tary of State for Education and Skills to the Institute for Public Policy Research. Retrieved November 18, 2008, from http://www.dfes.gov.uk/rsgateway/DB/STA/t000657/index.shtml.
de Abreu, G., & Cline, T. (2003). Schooled mathematics and cultural knowledge. *Pedagogy, Culture and Society*, 11(1), 11–30.
Delord, R., Vinrich, G., & Bourdais, M. (1996). *Cinq sur Cinq- Math 6ème*. Paris: Hachette Education.
Desrosières, A. (1998). *The politics of large numbers: A history of statistical reasoning*. Cambridge, MA: Harvard University Press.
Department for Education and Skills. (1999). *Framework for teaching mathematics: Reception to year 6*. London: Author.
Department for Education and Skills. (2001). *Key Stage 3 National Strategy—Framework for teaching mathematics: Years 7, 8 and 9*. London: Author.
Department for Education and Skills. (2003). Secondary Schools Curriculum and Staffing Survey November 2002 SFR 25/2003. Retrieved February 7, 2006, from http://www.dfes.gov.uk/rsgateway/DB/SFR/s000413/sfr25-2003.pdf
Department for Education and Skills. (2005). *The effects of pupil grouping: Literature review*. London: Author.
Dowling, P. (1996). A sociological analysis of school mathematics texts. *Educational Studies in Mathematics*, 31(4), 389–415.
Dowling, P. (1998). *The sociology of maths education: Mathematical myths/pedagogic texts*. London: Falmer.
Doyle, W. (1988). Work in mathematics classes: The content of student thinking during instruction. *Educational Psychologist*, 23(2), 167–180.
Drake, P. (2001). Mathematics and all that: Who teaches the number stuff? *Active Learning in Higher Education*, 2(1), 46–52.
Drake, P. (2005). A case of learning mathematics the hard way as a teaching assistant. *Research in Mathematics Education*, 7, 19–32.
Drake, P. (2006). *Working for learning: A case of mathematics for teaching*. Unpublished master's thesis, University of Sussex, Brighton.
Edwards, D., & Potter, J. (1992). *Discursive psychology*. Newbury Park, CA: Sage Publications.
Engestrom, Y., & Cole, M. (1997). Situated cognition in search of an agenda. In D. Kirshner & J. A. Whitson (Eds.), *Situated cognition: Social, semiotic and psychological perspectives* (pp. 301–309). Mahwah, NJ, London: Lawrence Erlbaum Associates.
Entwistle, N. J. (1981). *Styles of learning and teaching*. Chichester: John Wiley and Sons.
Evans, J. (2000). *Adults' mathematical thinking and emotions: A study of numerate practices*. London: RoutledgeFalmer.
Fairclough, N. (2003). *Analysing discourse: Textual analysis for social research*. London: Routledge.
Fine, M. (1998). Working in the hyphens: Reinventing self and other in qualitative research. In N. Denzin & Y. Lincoln (Eds.), *The landscape of qualitative research: Theories and issues* (pp. 130–155). Thousand Oaks, CA: Sage.
Finlow-Bates, K. (1997). *Investigating notions of proof: A study of students' proof activities within the context of a fallibilist and social theory*. Unpublished doctoral dissertation, London South Bank University, London, UK.
Foucault, M. (1972). *The archaeology of knowledge* (A. M. Sheridan Smith, Trans.). London: Routledge.
Foucault, M. (1980). Prison talk (C. Gordon, L. Marshall, J. Mepham, & K. Soper, Trans.). In C. Gordon (Ed.), *Power/knowledge* (pp. 37–54). Harlow: Prentice Hall.
Fox, J. (Ed.). (1987). *The essential Moreno: Writings on psychodrama, group method and spontaneity*. London: Springer.

Francis, B., & Archer, L. (2005). British-Chinese pupils' constructions of gender and learning. *Oxford Review of Education, 31*(3), 497–515.
Freudenthal, H. (1991). *Revisiting mathematics education: China lectures*. Dordrecht: Kluwer Academic Publishers.
Gardner, J. (2006). *Assessment and learning*. London: Sage.
Gardner, J. (2007, Summer). Is teaching a 'partial' profession? *Make the Grade*, pp. 18–21.
Gates, P., & Vistro-Yu, C. (2003). Is mathematics for all? In A. J. Bishop, M. A. Clements, C. Keitel, J. Kilpatrick, & F. K. S. Leung (Eds.), *Second international handbook of mathematics education Part One* (pp. 31–73). Dordrecht: Kluwer Academic Publishers.
Gee, J. P. (1999). *An introduction to discourse analysis: Theory and method*. London: Routledge.
Gee, J. P. (2000–2001). Identity as an analytic lens for research in education. *Review of Research in Education, 25*, 99–125.
Geertz, C. (1973). Deep play: Notes on the Balinese Cockfight. In C. Geertz (Ed.), *The interpretation of culture* (pp. 412–453). New York: Basic Books.
Gergen, K. (1999). *An invitation to social construction*. London: Sage.
Gilborn, D., & Youdell, D. (2000). *Rationing education: Policy, practice, reform, and equity*. Buckingham: Open University Press.
Goodchild, S., & Jaworski, B. (2005, July). *Identifying contradictions in a teaching and learning development project*. Paper presented at the 29th conference of the International Group for the Psychology of Mathematics Education, University of Melbourne, Australia.
Goulding, M., Rowland, T., & Barber, P. (2002). Does it matter? Primary teacher trainees' subject knowledge in mathematics. *British Educational Research Journal, 28*(5), 689–704.
Graven, M. (2002). Investigating mathematics teacher learning within an in-service community of practice: The centrality of confidence. *Educational Studies in Mathematics, 57*(2), 177–211.
Guardian. (2003). *Maths classes suffer teacher skills shortage*. Retrieved November 18, 2008, from http://www.guardian.co.uk/education/2003/sep/25/schools.uk1
Haggarty, L., & Pepin, B. (2002). An investigation of mathematics textbooks and their use in English, French and German classrooms: Who gets an opportunity to learn what? *British Educational Research Journal, 28*(4), 567–590.
Hall, S. (1991). Old and new identities, old and new ethnicities. In A. D. King (Ed.), *Culture, globalization and the world-system* (pp. 41–68). London: Macmillan.
Hall, S. (1996). Introduction: Who needs 'identity'? In S. Hall & P. du Gay (Eds.), *Questions of cultural identity* (pp. 1–17). London: Sage.
Halliday, M. A. K. (1985). *An introduction to functional grammar*. London: Edward Arnold.
Halliday, M. A. K. (1987). Language and the order of nature. In N. Fabb, D. Attridge, A. Durant, & C. McCabe (Eds.), *The linguistics of writing, arguments between language and literature* (pp. 135–154). New York: Methuen.
Halliday, M. A. K., & Hasan, R. (1985). *Language, context, and text: Aspects of language in a social-semiotic perspective*. Geelong, Australia: Deakin University.
Hamilton Maths Project/National Numeracy Project. (1998). Numeracy in action: Effective strategies for teaching numeracy [Video]. Oxford, UK: Author.
Hanley, U. (2007). Fantasies of teaching: Handling the paradoxes inherent in models of practice. *British Educational Research Journal, 33*(2), 253–271.
Hanley, U., & Jones, L. (2007). Mathematics and fantasies of effective practice. *Contemporary Issues in Early Childhood, 8*(1), 61–72.

Hardy, T. (2007). Participation and performance: Keys to confident learning in mathematics? *Research in Mathematics Education*, 9(1), 21–32.
Harford, T. (2007). *Waiting for good Joe*. Retrieved November 15, 2007, from http://www.slate.com/id/2177697/nav/tap3.
Harré, R., & Gillett, G. (1995). *The discursive mind*. Thousand Oaks, CA: Sage Publications.
Hasan, R. (2004). Ways of meaning, ways of learning: Code as an explanatory concept. *British Journal of Sociology of Education*, 23(4), 537–548.
Heidegger, M. (2000/1926). *Being and time* (J. Macquarrie & E. Robinson, Trans.). Oxford: Blackwell.
Heywood , D. (2007). Problematizing science subject matter knowledge as a legitimate enterprise in primary teacher education. *Cambridge Journal of Education*, 37(4), 519–542.
Hicks, D. (2002). *Reading lives: Working-class children and literacy learning*. New York: Teachers College Press.
Hiebert, J., Carpenter, T. P., Fennema, E., Fuson, K. C., Wearne, D., Murray, H., et al. (1997). *Making sense—teaching and learning mathematics with understanding*. Portsmouth, NH: Heinemann.
Hodgen, J., & Johnson, D. C. (2004). Teacher reflection, identity and belief change in the context of Primary CAME. In A. Millett, M. Brown & M. Askew (Eds.), *Primary mathematics and the developing professioanl* (pp. 219–244). Dordrecht: Kluwer.
Hodgen, J., & Askew, M. (2007). Emotion, indentity and teacher learning: becoming a primary mathematics teacher. *Oxford Review of Education*, 33(4), 469–487.
Holland, D., Lachicotte, W., Jr., Skinner, D., & Cain, C. (1998). *Identity and agency in cultural worlds*. Cambridge, MA: Harvard University Press.
Hollway, W., & Jefferson, T. (2000). *Doing qualitative research differently: Free association, narrative and the interview method*. London: Sage.
Holton, D. (Ed.). (2001). *The teaching and learning of mathematics at university level—An ICMI study* Dordrecht: Kluwer Academic Publishers.
Horn, I. S. (2007). Fast kids, slow kids, lazy kids: Framing the mismatch problem in mathematics teachers' conversations. *Journal of the Learning Sciences*, 16(1), 37–79.
Hoyles, C. (1982). The pupil's view of mathematics learning. *Educational Studies in Mathematics*, 13, 349–372.
Hoyles, C., Newman, K., & Noss, R. (2001). Changing patterns of transition from school to university mathematics. *International Journal of Mathematical Education in Science and Technology*, 32(6), 829–845.
Hughes, M., & Greenhough, P. (2008). 'We do it a different way at my school': Mathematics homework as a site for tension and conflict. In A. Watson & P. Winbourne (Eds.), *New directions for situated cognition in mathematics education* (pp. 129–152). New York: Springer.
Hundeland, P. S., Erfjord, I., Grevholm, B., & Breiteig, T. (2007). *Teachers and researchers inquiring into mathematics teaching and learning: The case of linear functions*. Paper presented at the Norma05, Fourth Nordic Conference on Mathematics Education, Trondheim, Norway.
IEA. (2005). *TIMSS 2003 International Report on Achievement in the Mathematics Cognitive Domains*. Retrieved November 10, 2008, 2007, from http://timss.bc.edu/timss2003i/mcgdm.html
Ireson, J., & Hallam, S. (1999). Raising standards: Is ability grouping the answer? *Oxford Review of Education*, 25(3), 343–358.
Ireson, J., Hallam, S., & Hurley, C. (2005). What are the effects of ability grouping on GCSE attainment? *British Educational Research Journal*, 31(4), 443–458.

Jaworski, B. (2003). Research practice into/influencing mathematics teaching and learning development: Towards a theoretical framework based on co-learning partnerships. *Educational Studies in Mathematics, 54*(2–3), 249–282.

Jaworski, B. (2004, July). *Grappling with complexity: Co-learning in inquiry communities in mathematics teaching development.* Paper presented at the 28th conference of the International Group for the Psychology of Mathematics Education, Bergen University College, Norway.

Jaworski, B. (2005). Learning communities in mathematics: Creating an inquiry community between teachers and didacticians. *Research in Mathematics Education, 7,* 101–119.

Jaworski, B. (2006). Theory and practice in mathematics teaching development: Critical inquiry as a mode of learning in teaching. *Journal of Mathematics Teacher Education, 9*(2), 187–211.

Jaworski, B., Fuglestad, A. B., Bjuland, R., Breiteig, T., Goodchild, S., & Grevholm, B. (Eds.). (2007). *Learning communities in mathematics.* Bergen: Caspar.

Jeffrey, B., & Woods, P. (1998). *Testing teachers: The effects of school inspections on primary teachers.* London: Falmer.

Jones, S., & Myhill, D. (2004). Seeing things differently: Teachers' constructions of underachievement. *Gender and Education, 16*(4), 531–546.

Jordan, N. (1968). *Themes in speculative psychology.* London: Tavistock.

Jørgensen, K. O., & Goodchild, S. (2007). Å utvikle barns forståelse av matematikk [To develop children's understanding of mathematics]. *Tangenten, 1,* 35–40, 49.

Kanes, C., & Lerman, S. (2008). Analysing concepts of community of practice. In A. Watson & P. Winbourne (Eds.), *New directions for situated cognition in mathematics education* (pp. 310–326). New York: Springer.

Kehily, M. J. (2001). Issues of gender and sexuality in schools. In B. Francis & C. Skelton (Eds.), *Investigating gender: Contemporary perspectives in education* (pp. 116–125). Buckingham: Open University Press.

Kendrick, M., & McKay, R. (2004). Uncovering literacy narratives through children's drawings. *Canadian Journal of Education, 27*(1), 45–60.

Kilpatrick, J., Swafford, J., & Findell, B. (Eds.). (2001). *Adding it up—Helping children learn mathematics.* Washington, DC: National Academy Press.

Kirshner, D. (2002). Untangling teachers' diverse aspirations for student learning: A crossdisciplinary strategy for relating psychological theory to pedagogical practice. *Journal for Research in Mathematics Education, 33*(1), 46–58.

Klein, M. (1935). A contribution to the psychogenesis of manic-depressive states. In *Love, guilt and reparation and other works* (pp. 263–289). London: Hogarth.

Klein, M. (1940). Mourning and its relation to manic-depressive states. In *Love, guilt and reparation and other works* (pp. 344–369). London: Hogarth.

Klein, M. (1946). Notes on some schizoid mechanisms. In *Envy and gratitude* (pp. 1–24). London: Hogarth.

Klein, M. (1959). Our adult world and its roots in infancy. In *Envy and gratitude and other works* (pp. 247–267). London: Hogarth.

Kring, T. (Writer). (2006). *Heroes.* Los Angeles, CA: NBC.

Kyriacou, C., & Goulding, M. (2004). *A systematic review of the impact of the Daily Mathematics Lesson in enhancing pupil confidence and competence in early mathematics.* London: EPPI-Centre.

Laity, P. (2007, November 5). The dangerous Don. *The Guardian,* p. 11.

Lakatos, I. (1978). *Proofs and refutations: The logic of mathematical discovery.* Cambridge: Cambridge University Press.

Lauder, H., Robinson, T., & Thrupp, M. (2002). School composition and peer effects. *International Journal of Educational Research, 37*(5), 483–504.

Lave, J. (1988). *Cognition in practice.* New York: Cambridge University Press.
Lave, J. (1996). Teaching as learning, in practice. *Mind Culture and Activity, 3*(3), 149–164.
Lave, J., & Wenger, E. (1991). *Situated learning: Legitimate peripheral participation.* Cambridge: Cambridge University Press.
Lerman, S. (1998). Cultural perspective on mathematics and mathematics teaching and learning. In F. Seeger, J. Voigt, & U. Waschescio (Eds.), *The culture of the mathematics classroom* (pp. 290–307). Cambridge: Cambridge University Press.
Lerman, S. (2001). Cultural, discursive psychology: A sociocultural approach to studying the teaching and learning of mathematics. *Educational Studies in Mathematics, 46,* 87–113.
Lerman, S. (2006). Learning mathematics as developing identity in the classroom. In P. Liljedhal (Ed.), *Proceedings of the Canadian Mathematics Education Study Group* (pp. 3–13). Ottawa: University of Ottawa, CMESG.
Lerman, S., & Tsatsaroni, A. (1998, September). *Why children fail and what mathematics education studies can do about it. The contribution/role of sociology.* Paper presented at the First International Mathematics Education and Society Conference, Nottingham, UK.
Lerman, S., & Zevenbergen, R. (2004). The socio-political context of the mathematics classroom: Using Bernstein's Theoretical Framework to understand classroom communications. In P. Valero & R. Zevenbergen (Eds.), *Researching the socio-political dimensions of mathematics education: Issues of power in theory and methodology* (pp. 27–42). Dordrecht: Kluwer.
Lo, J. (2000). Happy hybridity: Performing Asian-Australian identities. In I. Ang, S. Chalmers, L. Law, & M. Thomas (Eds.), *Alter/Asian* (pp. 152–168). Melbourne: Pluto Press.
Lucey, H., Melody, J., & Walkerdine, V. (2003). Uneasy hybrids: Psychosocial aspects of becoming educationally successful for working-class young women. *Gender and Education, 15*(3), 285–299.
Maasz, J., & Schloeglmann, W. (Eds.). (2006). *New mathematics education research and practice.* Rotterdam: Sense.
MacIntyre, H., & Ireson, J. (2002). Within-class ability grouping: Placement of pupils in groups and self-concept. *British Educational Research Journal, 28*(2), 249–263.
MacLure, M. (2003). *Discourse in educational and social research.* Buckingham: Open University Press.
Macrae, S., & Maguire, M. (2002). Getting in and getting on: Choosing the 'best'. In A. Hayton & A. Pacazuska (Eds.), *Access, participation and higher education: Policy and practice* (pp. 23–39). London: Kogan Page.
Marx, K. (1859/1970). *A contribution to the critique of political economy.* Moscow: Progree.
McIntyre, E., Roseberry, A., & Gonzalez, N. (Eds.). (2001). *Classroom diversity: Connecting curriculum to students lives.* Portsmouth, NH: Heinemann.
McLeish, J. (1991). *Number.* London: Bloomsbury.
McNamara, O., Jaworski, B., Rowland, T., Hodgen, J., & Prestage, S. (2002). *Developing mathematics teachers and teaching: A Research Monograph.* Retrieved October 26, 2007, from http://maths-ed.org.uk/mathsteachdev/pdf/mathsdev.pdf
Mendick, H. (2002, July). *'Why are we doing this?': A case study of motivational practices in mathematics classes.* Paper presented at the 26th conference of the International Group for the Psychology of Mathematics Education, Norwich, England.
Mendick, H. (2003a). Choosing maths/doing gender; a look at why there are more boys than girls in advanced mathematics classes in England. In L. Burton (Ed.),

Which way social justice for mathematics education (pp. 169–187)? Westport, CT: Praeger.
Mendick, H. (2003b). *Telling choices: an exploration of the gender imbalance in participation in advanced mathematics courses in England.* Unpublished doctoral thesis, Goldsmiths College, London.
Mendick, H. (2005a). A beautiful myth? The gendering of being/doing 'good at maths'. *Gender and Education, 17*(2), 89–105.
Mendick, H. (2005b). Mathematical stories: Why do more boys than girls choose to study mathematics at AS-level in England? *British Journal of Sociology of Education, 26*(2), 225–241.
Mendick, H. (2006). *Masculinities in mathematics.* Maidenhead: Open University Press.
Mitchell, J. (1986). *The selected Melanie Klein.* London: Penguin Books.
Mitchell, J. C. (1984). Typicality and the case study. In R. F. Ellen (Ed.), *Ethnographic research: A guide to general conduct* (pp. 238–241). London: Academic Press.
Moll, L., & Gonzalez, N. (1994). Lessons from research with language-minority children. *Journal of Reading Behavior, 26*(4), 439–456.
Moon, B. (1986). *The 'New Maths' curriculum controversy: An international story.* Lewes: Falmer.
Moreau, M.-P., Mendick, H., & Epstein, D. (2009). 'Terrified, mortified, petrified, stupefied by you' ... and gendered? Constructions of 'mathematical man' in popular culture. In E. Watson (Ed.), *Pimps, Wimps, Studs, Thugs and Gentleman: Essays on Media Images of Masculinity.* Jefferson: McFarland Publishers.
Morgan, C. (2009). Making sense of curriculum innovation and mathematics teacher identity. In C. Kanes (Ed.), *Developing professional practice in mathematics education.* New York: Springer.
Morgan, C. (2005). Words, definitions and concepts in discourses of mathematics, teaching and learning. *Language and Education, 19*(2), 103–117.
Morgan, C. (2006). What does social semiotics have to offer mathematics education research? *Educational Studies in Mathematics, 61*(1/2), 219–245.
Morgan, C., Tsatsaroni, A., & Lerman, S. (2002). Mathematics teachers' positions and practices in discourses of assessment. *British Journal of Sociology of Education, 23*(3), 445–461.
Moses, R. P., & Cobb, C. E. (2001). *Radical equations: Civil rights from Mississippi to the Algebra Project.* Boston: Beacon Press Books.
Mullis, I. V. S., Martin, M. O., Gonzalez, E. J., & Chrostowski, S. J. (2004). *TIMSS 2003 international mathematics report: Findings from IEA's Trends in International Mathematics and science study at the fourth and eighth grades.* Boston, MA: TIMSS & PIRLS International Study Center, Boston College.
Nardi, E., & Steward, S. (2003). Is mathematics T.I.R.E.D? A profile of quiet disaffection in the secondary mathematics classroom. *British Educational Research Journal, 29*(3), 345–367.
Nasir, N. S., & Cobb, P. (2002). Diversity, equity, and mathematical learning. *Mathematical Thinking and Learning, 4*(2–3), 91–102.
National Centre for Excellence in the Teaching of Mathematics. (2007). *Mathematics subject knowledge audits.* Retrieved November 6, 2007, from http://www.ncetm.org.uk/Default.aspx?page=20&module=mathematics_subject_knowledge_audits.
National Council of Teachers of Mathematics. (2000). *Principles and standards for school mathematics.* Reston, VA: National Council of Teachers of Mathematics/.
Nimier, J. (1993). Defence mechanisms against mathematics. *For the Learning of Mathematics, 13*(1), 30–34.

Nolan, K., & De Freitas, E. (Eds.). (2008). *Opening the research text: Critical insights and in(ter)ventions into mathematics education.* New York: Springer.
Nunes, T., Schliemann, A. D., & Carraher, D. W. (1993). *Street mathematics and school mathematics.* Cambridge: Cambridge University Press.
Ofsted. (1998). *Setting in primary schools.* London: Office for Standards in Education.
Ofsted. (2005). *New report shows rise in literacy and numeracy standards at Key Stage 2 after a four year plateau.* Retrieved October 26, 2007, from http://www.ofsted.gov.uk/portal/site/Internet/menuitem.e11147abaed5f711828a0d8308c08a0c/?vgnextoid=c30241fa2294c010VgnVCM1000003507640aRCRD.
Okri, B. (1983). *A way of being free.* London: Phoenix.
Organisation for Economic Co-operation and Development. (2004). *Learning for tomorrow's world: First results from PISA 2003.* Paris: OECD Publications.
Pampaka, M., Williams, J., Davis, P., & Wake, G. (2008, March). *Measuring pedagogic practice: A measure of 'teacher-centrism'.* Paper presented at the American Educational Research Conference, New York.
Parker, I. (2004). Psychoanalysis and critical psychology. In D. Hook, A. Collins, E. Burman, P. Kiguwa, & N. Mkhize (Eds.), *Critical psychology* (pp. 138–161). Cape Town: UCT Press.
Pepin, B. (1997). *Developing an understanding of mathematics teachers in England, France and Germany: An ethnographic study.* Unpublished doctoral thesis, Reading University, Reading, UK.
Pepin, B. (1999). The influence of national cultural traditions on pedagogy: Classroom practices in England, France and Germany. In J. Leach & B. Moon (Eds.), *Learners and pedagogy* (pp. 158–175). London: Sage.
Pepin, B. (2002). Methodological issues of cross-national comparisons: Efforts to establish equivalence in a cross-national study of mathematics teachers' work in England, France and Germany In A. Fries, M. Rosenmund, & W. Heller (Eds.), *Comparing curriculum making processes* (pp. 269–280). Zurich: Peter Lang.
Pepin, B., & Haggarty, L. (2001). Mathematics textbooks and their use in English, French and German classrooms: A way to understand teaching and learning cultures. *Zentralblatt for the Didactics of Mathematics, 33* (5), 158–175.
Pepin, B., & Haggarty, L. (2003). Mathematics textbooks and their use by teachers: A window into the education world of particular countries? In J. Van den Akker, W. Kuiper, & U. Hameyer (Eds.), *Curriculum landscapes and trends* (pp. 73–100). Dordrecht: Kluwer.
Pepin, B., & Haggarty, L. (2007, April). *Making connections and seeking understanding: Mathematical tasks in English, French and German textbooks.* Paper presented at the American Educational Research Association, Chicago, IL.
Picker, S. H., & Berry, J. S. (2000). Investigating pupils' images of mathematicians. *Educational Studies in Mathematics, 43*(1), 65–94.
Pickering, A. (1995). *The mangle of practice: Time, agency and science.* Chicago: Chicago University Press.
Pignatelli, F. (1993). What can I do? Foucault on freedom and the question of teacher agency. *Educational Theory, 43*(4), 411–432.
Popkewitz, T. S. (2004). School subjects, the politics of knowledge, and the projects of intellectuals in change. In P. Valero & R. Zevenbergen (Eds.), *Researching the socio-political dimensions of mathematics education: Issues of power in theory and methodology* (pp. 251–267). Boston: Kluwer Academic Publishers.
Povey, H., & Angier, C., with Clarke, C. (2006). Storying Joanne, an undergraduate mathematician. *Gender and Education, 18*(5), 459–471.
Povey, H., Boylan, M., Elliott, S., & Stephenson, K. (2000). 'An outrageous requirement'? Some responses from initial teacher education students to the imposition

of the Numeracy Skills Tests. *Proceedings of British Society for Research into the Learning of Mathematics Day Conference,* 20(2), 135–140.
Reay, D. (2002). Shaun's story: Troubling dominant discourses of working class masculinities. *Gender and Education, 14* (3), 221–234.
Reay, D., & Wiliam, D. (1999). 'I'll be a nothing': Structure, agency and the construction of identity through assessment. *British Educational Research Journal, 25*(3), 343–354.
Rocher, G. (1978). *Introducción a la sociología general.* Barcelona: Herder.
Rodd, M. (2005). Inclusion in practice: Special needs pupils in mainstream mathematics. In S. Johnstone-Wilder, P. Johnstone-Wilder, D. Pimm, & J. Westwell (Eds.), *Learning to teach mathematics in the secondary school* (pp. 187–205). London: Routledge.
Rodd, M., & Bartholomew, H. (2006). Invisible and special: Young women's experiences as undergraduate mathematics students. *Gender and Education, 18*(1), 35–50.
Rogoff, B., Matusov, E., & White, C. (1996). Models of teaching and learning: Participation in a community of learners. In D. Olsen and N. Torrance (Eds.), *The handbook of education and human development* (pp. 388–414). Oxford: Blackwell.
Rose, N. (1999). *Governing the soul* (2nd ed.). London: Free Association Books.
Roth, J. (1992). Of what help is he? A review of Foucault and Education. *American Educational Research Journal, 29*(4), 683–694.
Roth, W.-M., & McRobbie, C. J. (1999). Lifeworlds and the 'w/ri(gh)ting' of classroom research. *Journal of Curriculum Studies, 31*(5), 501–522.
Rotman, B. (1988). Towards a semiotics of mathematics. *Semiotica, 72*(1/2), 1–35.
Schmid, A., & Weidig, I. (1994). *Lambacher Schweizer 6.* Stuttgart: Klett.
Schmittau, J. (2003). Cultural-historical theory and mathematics education. In A. Kozulin, B. Gindis, V. S. Ageyev, & S. M. Miller (Eds.), *Vygotsky's educational theory in cultural context* (pp. 225–245). Cambridge: Cambridge University Press.
Schoenfield, A. (2002). Making mathematics work for all children: Issues of standards, testing, and equity. *Educational Researcher, 13*(1), 13–25.
Schoenfield, A. (2004). The math wars. *Educational Policy, 18*(1), 253–286.
Segal, H. (1973). *Introduction to the work of Melanie Klein.* London: Hogarth Press.
Seymour, E., & Hewitt, N. (1997). *Talking about leaving: Why undergraduates leave the sciences.* Boulder, CO: Westview Press.
Sfard, A. (1998). On two metaphors for learning and the dangers of choosing just one. *Educational Researcher, 27*(2), 4–13.
Sfard, A. (2007a). *Thinking as communicating: Human development, the growth of discourses, and mathematizing.* Cambridge: Cambridge University Press.
Sfard, A. (2007b). When the rules of discourse change, but nobody tells you: Making sense of mathematics learning from a commognitive standpoint. *Journal for Learning Sciences, 16*(4), 567–615.
Sfard, A. (2008). *Thinking as communicating: Human development, the growth of discourses, and mathematizing.* Cambridge: Cambridge University Press.
Sfard, A., & Prusak, A. (2005a, July). *Identity that makes a difference: Substantial learning as closing the gap between actual and designated identities.* Paper presented at the 29th meeting of the International Group for the Psychology of Mathematics Education, University of Melbourne, Victoria, Australia.
Sfard, A., & Prusak, A. (2005b). Telling identities: In search of an analytic tool for investigating learning as a culturally shaped activity. *Educational Researcher, 34*(4), 14–22.
Shaw, J. (1995). *Education, gender and anxiety.* London: Taylor & Francis.
Skeggs, B. (2004). *Class, self, culture.* London: Routledge.

Smith, A. (2004). *Making mathematics count*. London: Department for Education and Skills.
Smith, M. P., & Stein, M. K. (1998). Selecting and creating mathematical tasks: From research to practice. *Mathematics in the Middle School, 3*(5), 344–350.
Smithers, A. (2003). *Study for DfES reported in BBC News 'Teachers quit over reform overload'*. Retrieved November 11, 2007, from http://news.bbc.co.uk/1/hi/educationw/3020106.stm.
Solomon, Y. (2006). Deficit or difference? The role of students' epistemologies of mathematics in their interactions with proof. *Educational Studies in Mathematics, 61*(3), 373–393.
Solomon, Y. (2007a). Experiencing mathematics classes: Gender, ability and the selective development of participative identities. *International Journal of Educational Research, 46*(1/2), 8–19.
Solomon, Y. (2007b). Not belonging? What makes a functional learner identity in the undergraduate mathematics community of practice? *Studies in Higher Education, 32*(1), 79–96.
Solomon, Y. (2007c, September). *Theorising fragile identities: What can we learn from mathematics students?* Paper presented at the Second Socio-Cultural Theory in Educational Research and Practice Conference, Manchester, UK.
Solomon, Y. (2008). *Mathematical literacy: Developing identities of inclusion*. London/New York: Routledge.
Spradley, J. P. (1979). *The ethnographic interview*. New York: Holt, Rinehart & Winston Ltd.
Stronach, I., & MacLure, M. (1997). *Educational research undone: The postmodern embrace*. Buckingham: Open University Press.
Suggate, J., Davies, A., & Goulding, M. (2006). *Mathematical knowledge for primary teachers* (3rd rev. ed.). Abingdon: David Fulton Publishers Ltd.
Sukhnandan, L., & Lee, B. (1988). *Streaming, setting and grouping by ability: A review of the literature*. Slough: National Foundation for Educational Research.
Suurtamm, C., & Vezina, N. (2004, July). *Longitudinal study of professional development to build primary teacher expertise in teaching mathematics*. Paper presented at the annual meeting of North American chapter of the International Group for the Psychology of Mathematics Education, Toronto, Canada.
Symington, J., & Symington, N. (2004/1996). *The clinical thinking of Wilfred Bion*. London: Routledge.
TDA. (2007). *Maths pilot CPD programme*. Retrieved October 11, 2007, from http://www.tda.gov.uk/teachers/continuingprofessionaldevelopment/maths_cpd.aspx
Teachernet. (2007). *Study materials: Auditing your own knowledge of mathematics*. Retrieved November 6, 2007, from http://www.teachernet.gov.uk.
Thrupp, M. (1997, March). *The school mix effect, how the social class composition of school intakes shapes school processes and student achievement*. Paper presented at the annual meeting of the American Educational Research Association. Chicago, IL.
Thrupp, M. (1998). Exploring the politics of blame: School inspection and its contestation in New Zealand and England. *Comparative Education, 34*(2), 195–209.
TIME Magazine. (2007). *Numbers*. Retrieved November 21, 2007, from http://www.time.com/time/magazine/article/0,9171,1668461,00.html?iid=digg_share.
Tsatsaroni, A., Lerman, S., & Xu, G. (2003, April). *A sociological description of changes in the intellectual field of mathematics education research: Implications for the identities of academics*. Paper presented at the annual meeting of the American Educational Research Association, Chicago, IL.
Urban Dictionary, http://www.urbandictionary.com/define.php?term==neek. Accessed 2 February 2009.

Valero, P. (2007). A socio-political look at equity in the school organization of mathematics education. *Zentralblatt für Didaktik der Mathematik, 39*(3), 225–233.
Valverde, G. A., Bianchi, L. J., Wolfe, R. G., Scmidt, W. H., & Houng, R. T. (2002). *According to the Book—Using TIMSS to investigate the translation of policy into practice through the world of textbooks.* Dordrecht: Kluwer Academic Publishers.
Van Manen, M. (1990). *Researching lived experience: Human science for an action sensitive pedagogy.* New York: State University of New York Press.
Vickers, K. M., Tipler, M. J., & van Hiele, H. L. (1996a). *New National Curriculum 8.* Cheltenham: Stanley Thornes.
Vickers, K. M., Tipler, M. J., & van Hiele, H. L. (1996b). *New National Curriculum Mathematics A/A*.* Cheltenham: Stanley Thornes.
von Glasersfeld, E. (1990). An exposition of constructivism: Why some like it radical. In R. B. Davis, C. A. Maher, & N. Noddings (Eds.), *Constructivist views on the teaching and learning of mathematics* (pp. 19–29). Reston VA: National Council of Teachers of Mathematics.
Vygotsky, L. S. (1962). *Thought and language* (A. Kozulin, Trans.). Cambridge, MA: The MIT Press.
Vygotsky, L. S. (1978). *Mind in society* (M. Cole & S. Scribner, Trans.). Cambridge, MA: Harvard University Press.
Vygotsky, L. S. (1987). Thinking and speech (N. Minick, Trans.). In R. W. Rieber & A. C. Carton (Eds.), *The collected works of L. S. Vygotsky* (pp. 39–285). New York: Plenum Press.
Wacquant, L. (1989). Towards a reflexive sociology: A workshop with Pierre Bourdieu. *Sociological Theory, 7*(1), 26–63.
Waddell, M. (1998). *Inside lives: Psychoanalysis and the growth of the personality.* London: Taylor & Francis.
Waddell, M. (2002). *Inside lives: Psychoanalysis and the growth of the personality.* London: Karnac.
Walkerdine, V. (1988). *The mastery of reason: Cognitive development and the production of rationality.* London: Routledge.
Walkerdine, V. (1997). Redefining the subject in situated cognition theory. In D. Kirshner & J. A. Whitson (Eds.), *Situated cognition: Social, semiotic and psychological perspectives* (pp. 57–70). Mahwah, NJ: Lawrence Erlbaum Associates.
Walkerdine, V. (1998). *Counting girls out* (2nd ed.). London: Falmer.
Walkerdine, V. (2007). *Children, gender, video games: Towards a relational approach to multimedia.* Basingstoke: Palgrave Macmillan.
Wake, G., Williams, J., Black, L., Davis, P., Hernandez-Martinez, P., & Pampaka, M. (2007, September). *Pedagogic practices and interweaving narratives in AS Mathematics classrooms.* Paper presented at the British Educational Research Association Conference, Institute of Education, University of London.
Wang, J., & Lin, E. (2005). Comparative studies on US and Chinese mathematics learning and the implications for standards-based mathematics teaching reform. *Educational Researcher, 34*(5), 1–13.
Warwick, D., & Osherson, S. (Eds.). (1973). *Comparative research methods: An overview.* Englewood Cliffs: Prentice-Hall.
Watkins, C., & Mortimore, P. (1999). Pedagogy: wWhat do we know? In P. Mortimore (Ed.), *Understanding pedagogy and its impact on learning* (pp. 1–20). London: Paul Chapman/Sage.
Watson, A. (2006). Some difficulties in informal assessment in mathematics. *Assessment in Education, 13*(3), 289–303.
Wenger, E. (1998). *Communities of practice: Learning, meaning and identity.* Cambridge: Cambridge University Press.

Wetherell, M. (2001). Debates in discourse research. In M. Wetherell, S. Taylor, & S. J. Yates (Eds.), *Discourse theory and practice: A reader* (pp. 380–399). London: Sage.
Williams, G. (1997). *Internal landscapes and foreign bodies. Eating disorders and other pathologies*. London: Duckworth.
Williams, J., Davis, P., & Black, L. (2007). Subjectivities in school: Socio-cultural and activity theory perspectives. *International Journal of Educational Research*, 2(1–2), 1–7.
Wilson, S., Winbourne, P., & Tomlin, A. (2008). 'No way is can't': A situated account of one woman's uses and experiences of mathematics. In A. Watson & P. Winbourne (Eds.), *New directions for situated cognition in mathematics education* (pp. 327–349). New York: Springer.
Winbourne, P. (2007, September). *Who's the researcher? Methodological difficulties of making and understanding personal accounts of experiences of mathematics education*. Paper presented at the Second Socio-cultural Theory in Educational Research and Practice Conference: Theory, Identity and Learning, Manchester University, Manchester, UK.
Winbourne, P. (2008). Looking for learning in practice: How can this inform teaching? In A. Watson & P. Winbourne (Eds.), *New directions for situated cognition in mathematics education* (pp. 79–102). New York: Springer.
Winbourne, P., & Watson, A. (1998a). Learning mathematics in local communities of practice. In A. Watson (Ed.), *Situated cognition in the learning of mathematics* (pp. 93–104). Oxford: University of Oxford, Centre for Mathematics Education Research.
Winbourne, P., & Watson, A. (1998b, July). *Learning mathematics in local communities of practice*. Paper presented at the 22nd annual meeting of the International Group for the Psychology of Mathematics Education, Stellenbosch, South Africa.
Winnicott, D. W. (1971). *Playing and reality*. London: Routledge.
Wittgenstein, L. (1953). *Philosophical investigations*. Oxford: Blackwell.
Woods, P. (1986). *Inside schools: Ethnography in educational research*. London: Routledge & Kegan Paul.
Yackel, E., Cobb, P., & Wood, T. (1999). The interactive constitution of mathematical meaning in one second grade classroom: An illustrative example. *Journal of Mathematical Behavior*, 17(4), 469–488.
Yalom, I. (1980). *Existential psychotherapy*. New York: Basic Books.
Zevenbergen, R. (1991, July). *Children's conception of space*. Paper presented at the annual Mathematics Education Research Group of Australasia Conference MERGA 14, Perth, Australia.
Zevenbergen, R. (2000). 'Cracking the code' of mathematics classrooms: School success as a function of linguistic, social and cultural background. In J. Boaler (Ed.), *Multiple perspectives on mathematics teaching and learning* (pp. 201–223). Westport, CT: Ablex Publishing.
Zevenbergen, R. (2001). Mathematics, social class, and linguistic capital: An analysis of mathematics classroom interactions. In B. Atweh, H. Forgasz, & B. Nebres (Eds.), *Sociocultural research on mathematics education* (pp. 201–215). Mahwah, NJ: Lawrence Erlbaum Associates.
Žižek, S. (2006). *How to read Lacan*. London: Granta Publications.

Contributors

Tamara Bibby is interested in psychosocial explorations of classrooms, particularly young people's understandings of their learning relationships and processes, group processes, and primary teachers' views of themselves and their professional roles. Mathematics continues to be of particular interest as a site for research into aspects of learning and relating.

Laura Black is Lecturer at the University of Manchester. Her research interests focus on pedagogic processes in the classroom and the construction of learner identities, particularly in relation to mathematics. She is currently working on several projects on widening participation in mathematically demanding programmes within postcompulsory education.

Mark Boylan is Senior Lecturer in Secondary Mathematics Education at Sheffield Hallam University. His research interests include social interaction in mathematics classrooms and emotional and personal development of teachers. He is an active researcher of his own practice, focussing on the development of learning communities and peer learning with Initial Teacher Training students.

Margaret Brown is Professor of Mathematics Education at King's College London. She has directed more than 25 research projects in the learning, teaching, and assessment of mathematics at all ages from early years to adult. She has been President of the British Educational Research Association and Chair of the Education panel for the UK's Research Assessment Exercise 2008.

Pauline Davis is Senior Lecturer in Education at the University of Manchester. She has led a number of research projects in mathematics education, inclusive education, and financial literacy for the Economic and Social Research Council, government, and others. She has a broad interest in research methodology in the context of sociocultural theories (CHAT) and inequality.

Pat Drake has three strands to her work currently: mathematics education, teacher education, and practitioner/insider research. She works at the University of Sussex, where she is a Senior Lecturer in Education.

Debbie Epstein is Professor of Education at Cardiff University. She has published widely on questions of gender, sexuality, and race and their intersections in educational sites and popular culture. She has written extensively on higher education. Together with Rebecca Boden and Jane Kenway, she is author of the *Academic's Support Kit*.

Patricia George has taught mathematics in secondary schools in Antigua & Barbuda, Caribbean, for 14 years. She recently completed her PhD at the University of Leeds; this study employed a sociocultural approach to an investigation of Caribbean students' relationships with mathematics. Presently, she works in the Ministry of Education, Antigua & Barbuda as a Senior Research Officer.

Tansy Hardy is Lecturer in Mathematics Education at Sheffield Hallam University and leads master's courses in Learning and Teaching. She is concerned with the inevitable political nature of both teaching and learning in mathematics. Her current research examines how this affects teachers' and learners' (and researchers') senses of themselves and of future possibilities.

Jeremy Hodgen is Senior Lecturer in Mathematics Education at King's College London. His research interests include assessment, teacher education, and the relationship between research and practice in school mathematics. He is currently investigating lower secondary students' understandings of algebra and multiplicative reasoning and how teachers can respond to this awareness.

Barbara Jaworski is currently Professor in Mathematics Education in the School of Mathematics at Loughborough University. She has worked extensively in teacher and researcher education in the UK and Norway. Her research primarily is into the development of mathematics teaching involving developmental research approaches and formation of inquiry communities between teachers and educators.

Steve Lerman is Professor of Mathematics Education at London South Bank University, where he runs the professional doctorate programme and is Head of Educational Research. His current research interests include sociocultural and sociological studies, equity, and the role of theoretical frameworks in research in the field. He is a former President of the International Group for the Psychology of Mathematics Education.

Rachel Marks has previously taught for 5 years in primary schools and is currently an Economic and Social Research Council-funded PhD student at King's College London. Her research explores the production, reproduction, and contestation of ability in primary school mathematics from the perspectives of both pupils and teachers.

Heather Mendick is Senior Research Fellow at the Institute for Policy Studies in Education, London Metropolitan University, and lecturer at Goldsmiths, University of London. In her research, she works across education, gender studies, sociology, and cultural studies, and she is particularly interested in influences of popular culture on identities and aspirations. She used to be a mathematics teacher.

Marie-Pierre Moreau is Research Fellow at the Institute for Policy Studies in Education, London Metropolitan University. Her current research interests are education and employment policies in relation to social justice, with a particular focus on gender and its discursive construction. Areas she has researched recently include: graduates' employment, teachers' careers, school equal opportunities policies, and people's relationship with mathematics.

Candia Morgan is Reader in Mathematics Education at the Institute of Education, University of London. Before working in higher education, she taught mathematics in London secondary schools, and it is from this experience that her interest in language and discourse arose. Her research adopts a critical perspective on discourses of mathematics, teaching, and curriculum.

Birgit Pepin is Professor of Mathematics Education at the University College in Trondheim, Norway, having previously worked at the University of Manchester. Her main research interests are to explore and compare issues in mathematics classrooms internationally (funded by the Economic and Social Research Council [ESRC] and the European Union). Recent research, also funded by the ESRC, investigates transition practices from college/school into higher education in mathematics.

Hilary Povey is Professor of Mathematics Education at Sheffield Hallam University, where she teaches mathematics at the undergraduate level to initial teacher education students. Throughout her career, in teaching, curriculum development, and research, she has had a commitment to promoting social justice in and through the mathematics classroom.

Melissa Rodd works at the Institute of Education, University of London. She is a member of the Mathematics Special Interest (research and development) Group and teaches undergraduate, master's, and doctoral students in mathematics education. Her central interest is in mathematical experience and its expression in artefacts, emotions, and communities.

Anna Sfard's research focuses on the development of mathematical discourses in individual lives and in the course of history. She is the author of *Thinking as Communicating: Human Development, the Growth of Discourses, and Mathematizing* and the recipient of the 2007 Freudenthal Medal for research in mathematics education.

Jenny Shaw is currently completing a book on shopping and working on a biographic study of the 'Urban Youth', the teenagers sent to the countryside by Mao Tse Tung. She has worked on gender in education and the workplace, on life histories, and using Mass Observation. Her approach tends toward the psychosocial.

Yvette Solomon is Professor of Education at Manchester Metropolitan University. Her main interest is in the development of mathematical identities in secondary and undergraduate mathematics students, focusing on the nature of relationships among language, learning, and mathematical knowledge, as well as among identity, participation, and community experience.

Paola Valero is Associate Professor in Mathematics Education at the Department of Education, Learning and Philosophy, Aalborg University, Denmark. Her research interests are, among others, the political dimension of mathematics education in areas such as school reform processes, curricular innovation, and multiculturalism and mathematics learning.

Julian Williams is Professor of Mathematics Education and Director of Post Graduate Research studies in the School of Education, University of Manchester. He is currently leading or involved in a number of research projects investigating mathematics learners' in transition, but has a broad interest in sociocultural and cultural-historical activity theory.

Peter Winbourne is Reader in Educational Development at London South Bank University, UK. He was a teacher of mathematics and advisory teacher in London schools for 18 years before moving into higher education in the early 1990s. His main research focus is the development and application of theories of situated cognition and identity.

Index

Page numbers in *italics* indicate tables.

A

ability grouping 58–9, 68–9, 110, 117, 181, 203; ethics, limitations, and focus 59–60; parent perspectives 58–9, 61–3, 69; pupil perspectives 65–8; study methodology 63–5; 'talented and gifted' 66–7; theoretical perspective 60–1; and use of textbooks 102–5, 114; *see also* identity and ability

Adler, J. 152, 158, 196

alignment, in a community of practice 174, 176, 177, 183

anxiety 87–8; feeding analogy of learning 94–5; and feelings of inadequacy 25–6; and hatred of mathematics 49–51; people as numbers and numbers as people 91–3; 'shadow side' of mathematics 88–91

Apple, M. 211

assessment and selection 5–42, 203; numerical assessments 12, 15–18; relational approaches 41; *see also* defended subject; identity and ability (primary education); numberese

B

Bakhtin, M. 17
Ball, S.J. 3, 58, 71, 130, 150
Ball, D. L. 92, 190–191
Bartholomew, H. 3, 33, 48, 150
belong(ing) 10, 29, 132, 133, 143, 145, 158, 174–175, 177, 206, 217
Benjamin, J. 125
Benwell, B. & Stokoe, E. 98
Bernstein, B. 59, 97, 120, 136–7, 147–8, 151, 152, 153

Bibby, T. 3, 6, 39, 74, 95, 119, 150, 157, 192, 204, 206, 211, 223–224
Bion, W. 89, 93, 94–5, 125–6, 132
Black, L. 3, 144 et al. 6, 208, 213, 214, 223
Black, P. et al 41; & Wiliam, D. 30, 41
blame: and guilt 130–1, 134–5; of maths 54; primary school teaching problems 189–90; self- 44, 45, 76
Boaler, J. 1, 3, 41, 48, 69, 70, 109, 110, 136, 150; & Greeno, J.G. 2, 32, 33, 47, 48, 56, 108–9, 109, 136, 144, 149, 150; et al. 3, 48, 70, 108, 110
boundary crossing 141–2, 152–4
boundary maintenance 72, 79–80, 81
Bourdieu, P. 206, 209–210
Boylan, M. 48, 54, 57; & Povey, H. 43–44,157, 203, 207, 210, 212, 224
Britzman, D. 2, 123–4, 126
Brown, M. 1, 6, 208; et al. 32, 35, 180, 189
Brown, S. & MacIntyre, D. 176
Brown, T. 3; et al. 47
Burton, L. 57, 76, 155, 194
Butler, J. 78, 159
Buxton, L. 161, 162

C

Cane, A. 87
Canham, H. 91–2
Carr, M. 32
choice 43–82; 205–6; *see also* ability grouping; stories (primary school experience)

Chouliaraki, L. & Fairclough, N. 98, 105
classroom assistants' perspectives 169–70
classroom environment in different countries 110–11, 117–18
Cobb, P. 136, 144
collaborative inquiry (LCM project): growth and identity 183–4; issues 179–82; tensions and resolutions 182–3; theoretical perspective 174–7; as tool for learning 177–9
collaborative student relationships *see* hybridity of maths and peer talk
'commognition' 147
community of inquiry 173–184
community of practice 44, 61, 146, 174–7
confidence 192–4
consumerism, subject choices, and popular culture 74–6
context embeddedness of tasks 112–13
Cooper, B. 113, 152
critical discourse analysis (CDA) 98–9, 105
Cronin, A. 75
curriculum 83–118; 203–4, 209; knowledge *vs* process knowledge 126–7; *vs* inquiry-based approach 180–1; *see also* anxiety; curriculum discourse; textbooks, role in English, French, and German classrooms
curriculum discourse 97–8; analytical method 98–9; guidance for teachers 99–101; textbooks for GCSE students 101–5

D

Damarin, S. 79
Davis, P. 137, 144; & Williams, J. 119–120, 158, 211, 212, 223
defended subject 19–22; graduate perspectives 22–8; psychic power of assessment and selection 29–30
Department for Education and Skills 70, 98, 99–100, 112, 162–3
depressive position 20–1, 22, 24, 28
detraditionalisation 149
'doer/done to' teacher-pupil relationship 130–2, 134–5
Dowling, P. 3, 101–102, 104, 113
Drake, P. 157, 163, 164, 204–205, 207, 208, 210, 214

E

engagement 174, 181; and agency 102–5
enjoyment: identity and ability in subject choice 74–80; *see also* 'fun' of mathematics
Evans, J. 2

F

failure, sense of 89
Fairclough, N. 98–99; *see also* Chouliaraki, L. & Fairclough, N.
fashions in teaching 185–6
fathers, relationships with 24, 27–8, 98, 164, 169
feeding analogy of learning 94–5
'figured worlds' 108–9
Foucault, M. 72, 73, 187, 192, 194
Freudenthal, H. 111
'fun' of mathematics 39, 133, 138

G

Gardner, J. 40, 41
Gates, P. & Vistro-Yu, C. 209, 210
Gee, J.P. 32, 36, 60–1, 63–4
'geeks' 33, 79–80; and 'neeks' 65, 70
gender 3, 23, 44, 45, 48, 72, 76, 78, 81, 83, 143, 144, 149, 150, 165, 166–9, 215; maleness of mathematics 53, 89–90
generality of numerical assessments 12, 14–15
'gifted and talented' groups 44, 66–7
Goulding, M. et al. 191
graduate perspectives 22–8
grounded theory 34
group(s): interaction 154–5; *vs* individuals 129, 181–2; *see also* ability grouping
guilt 22, 28, 150; and blame 130–1, 134–5

H

Hall, S. 71, 148
Halliday, M.A.K. 13, 105, 144, 147
Hanley, U. 164
Hardy, T. 158, 193, 204–205, 210, 225
Harford, T. 10
hatred of mathematics 49–51, 74
Heywood, D. 196
Hicks, D. 143
Hiebert, J. et al. 108, 116–17
hierarchies 180
Hodgen, J. 35, 41; & Marks, R. 5, 7, 158, 203, 206, 209, 223

Holland, D. et al. 33, 61, 108–9, 201–2, 207
Hollway, W. & Jefferson, T. 19
hybridity of maths and peer talk 136–7, 145–6; background 137–8; boundary crossing 141–2; classroom pedagogic culture and identity 142–5; statistics class 138–41

I
idealisation 21, 124, 133, 134–5
identity 60–1, 98, 104–5, 144–5, 148, 208–11; and ability 31- 41; and ability and enjoyment 78; and subject choice 74–80; change of 26–7; concepts, definitions and theoretical perspectives 215–18; gendered 143; and growth 183–4; ideas of 201–6; impact of regulation on 150–1; importance of 206–8; in late capitalist era 149; manifestation of 149–50; in mathematical histories 34,37; multidimensionality of 137, 212; narratives of 10; negation of self 131; and neoliberalism 75–6; pedagogic culture and 142–5; school mathematics 151–5; and subjectivity 73; usefulness of concept 218–19; *see also* pedagogical practices impact on learner identities; stories; narratives
imagination 10, 20, 40, 158, 174
inchoate and unknown aspects of education 124, 126–7, 134, 135
individuals *vs* groups 129, 181–2
in(ex)clusion standpoints 222–6
inquiry: community of 173, 174; conceptual difficulties 180; as tool for learning 177–9; *see also* collaborative inquiry (LCM project)
Internet measuring tools 10–11
intersubjectivity 125, 126, 129–30, 135, 219–220

J
Jaworski, B. 177, 179, 182, 183, 204, 212,
Jeffrey, B. & Woods, P. 150

K
Kanes, C. & Lerman, S. 61

Klein, M. 20–2, 157
Knowledge Link (K Link) 125–7, 132–3, 134

L
labelling 5–7, 11, 34, 35, 36–7, 41, 44, 48, 55, 66, 161, 225
Lacan, J. 147, 157, 162,164–165, 170–1
Lave, J. 14, 149, 175–7; & Wenger, E. 2, 61, 108, 149, 174, 201
Lerman, S. 150, 158, 196, 203, 206; & Tsatsaroni, A. 148; & Zevenbergen, R. 148, 151 *see also* Kanes & Lerman
Lo, J. 142
local and national projects (primary school teaching) 190–4

M
Maclure, M. 48, 72, 185–6, 194
McNamara, O. et al. 191, 195, 196, 197
maleness of mathematics 53, 89–90, 215
'manic defences' 19–20
Marks, R. 5, 158, 203, 206, 209,
Marx, K. 148
mathematical tasks 107–8, 111–18, 154–55, 176, 178, 180
measuring 10, 16, 55–6, 126, 190
measurement 6, 7, 9, 11, 41, 110
Mendick, H. 6, 29, 45, 47, 48, 74, 76, 80, 150, 158, 186, 205, 208
Mitchell, J. 20
Mitchell, J.C. 34
Moreau, M-P. 29, 45, 80, 158, 205
Morgan, C. 83, 100, 104, 105, 106, 148, 158, 206
mother/infant relationship 21–22, 84, 89, 94–5, 96, 164
motivation of teachers 164–70, 171
multidimensionality of identities 212
mutuality 129, 132–3

N
narrative 10, 17, 19, 20, 54, 57, 157, 208, 224; analysis 34; identity 10, 145; *see also* stories
Nasir, N.S. 136
National Centre for Excellence in the Teaching of Mathematics 158–9, 193
national and local projects (primary school teaching) 190–4

negation of self 131
neoliberalism 71, 73, 75–6
Nimier, J. 19–20, 28, 29, 30
numberese 215, 225: educational uses of 12–17; hegemony of 9–11; limitations of 17–18
Nunes, T. 202

O
Ofsted 31, 189
'others'/'othering' 56–7, 79–80, 95, 164, 165

P
pain of learning 120, 126, 134, 171
pains and pleasures of mathematics 6, 23–30, 43, 57
paranoid-schizoid position 20–2, 28, 29
parents: fathers, relationships with 24, 27–8, 98, 164, 169; mother/infant relationship 21–22, 84, 89, 94–5, 96, 164; perspectives on ability grouping 58–9, 61–3, 69; relationship between 92
Parker, I. 191, 192
participation 100–1, 174–6, 193–4, 201–2, 203, 206–7, 222, 223
pedagogical practices impact on learner identities: cross-cultural 110–116; K Link in primary school 132–3, 134; mutuality 129, 132–3; psychoanalytical approach 123–7; relationship focus in secondary education 127–9; research contexts 127; valuing 129–32
pedagogy 2, 3, 63, 70, 82, 101, 119–21, 126, 136, 144, 145, 151–2, 155, 158, 191, 195, 196, 204, 220; definitions of 123, 127, 129, 133, 135, 223
Pepin, B. 110, 111, 115, 158, 203, 206; & Haggarty, L. 109, 111, 113, 114
phantasy 20–2, 26, 29
'phobic defences' 19–20
popular culture, subject choices and consumerism 74–6
Povey, H. 43, 48, 157, 207, 212; & Angier, C. with Clarke, C. 47, 48; & Boylan, M. Elliot, S. & Stephenson, K. 57
primary education: K Link in 132–3, 134; see also stories
primary school teaching problems 188–9; and blame 189–90; and local and national projects 190–4; and possible solutions 194–7
psychoanalytical approaches 19, 87–8, 89, 157, 164–6, 170–1 see also anxiety; defended subject; unconscious processes
pupil perspectives: ability grouping 65–8; identity and ability (primary education) 34–7; impact of pedagogical practices 129–32, 132–3, 134
qualifications 162–3; classroom assistants 169–70; graduate trainee teachers 166–9

R
Reay, D. 150; & Wiliam, D. 3, 34
regulation, in pedagogic relations, 147–8; impact on identity of 150–1
relational knowing 125–6 see also valuing in teacher-pupil relationships
Rodd, M. 3 ; & Bartholomew, H. 3
Rogoff, B. et al. 175
Rose, N. 71, 80–1
Roth, W.-M. 48
Roth, J. 195

S
Schoenfeld, A. 97, 108, 209
'Sean's numbers' 92–3
Segal, H. 20, 22
self see identity
setting see ability grouping
Sfard, A. 31, 40, 147, 150, 205, 209
'shadow side' of mathematics 88–91
Shaw, J. 3, 29, 96
Skeggs, B. 76
social class 3, 45, 72, 76, 81, 120, 143–4, 147, 149, 151–2
Solomon, Y. 3, 6, 29, 47, 145, 208, 214, 223
'splitting' 21–2, 23–4, 27–9; pains and pleasures of mathematics 23–8
stories 10, 17, 20, 29, 60 72, 74, 81–2; about mathematics 47–57; of compulsion and struggle 54–5; of measuring 55–6 see also identity; narratives
streaming see ability grouping

subjectivity: and identity 73, 125, 217; and knowledge 123; intersubjectivity 125, 126, 128–30, 135, 219
Suggate, J. et al. 189
Suurtamm, C. & Vezina, N. 190–1
symbolic control 148
symbolic order 164
symbolic other 164, 165
Symington, J. & Symington, N. 126

T

teacher development 204–5; see also collaborative inquiry (LCM Project); primary school teaching problems
teacher guidance, discourse of 99–101, 187–8
teacher perspectives 162–4; on identity and ability 37–40; motivation to teach 164–70, 171
Teachernet 192
television portrayal of mathematicians 80, 81
textbooks: ability levels 102–5, 114; discourse of 101–5; role in English, French, and German classrooms 107–18
Thrupp, M. 144
'truths', discursive approach to 186–8; mathematical 16, 17

U

unconscious processes 2, 88, 91, 123–7, 211; see also anxiety; defended subject; psychoanalytical approaches
undergraduates: choice 76–80; experiencing university mathematics 23–8; group interaction 154–5
unknown and inchoate aspects of education 124, 126–7, 134, 135

V

Valero, P. 214, 212, 223
valuing in pupil/teacher relationships 129–32
Vygotsky, L. 9, 124, 145, 147, 148, 175

W

Waddell, M. 20, 21, 22, 94
Watson, A. 41, 61, 151
Walkerdine, V. 3, 20, 23, 47, 56, 73, 77, 80
Watkins, C. & Mortimore, P. 123
Wenger, E. 31–2, 61, 137, 144, 146, 174–6, 177, 201, 206, 217; Lave, J. & 2, 108, 149, 174–6, 201
whole-class discussion 85, 111, 115, 137
Williams, J. et al. 137, 144, 145
Williams Review 194–5
Winbourne, P. 44, 61, 64, 69, 158, 210, 211, 224–5 ; et al. 61, 64, 151
Winnicott, D. 89

Y

Yackel et al. 136

Z

Zevenbergen, R. 3, 108, 113, 148, 151
Žižek, S. 165